An Account Of Thermodynamic Entropy

Authored by:

Alberto Gianinetti

Council for Agricultural Research and Economics
Genomics Research Centre, via San Protaso 302,
29017 Fiorenzuola d'Arda
Italy

An Account of Thermodynamic Entropy

Author : Alberto Gianinetti

eISBN (Online): 978-1-68108-393-3

ISBN (Print): 978-1-68108-394-0

First published in 2017.

advertisements or ideas contained in the Work.

Limitation of Liability:

In no event will Bentham Science Publishers, its staff, editors and/or authors, be liable for any damages, including, without limitation, special, incidental and/or consequential damages and/or damages for lost data and/or profits arising out of (whether directly or indirectly) the use or inability to use the Work. The entire liability of Bentham Science Publishers shall be limited to the amount actually paid by you for the Work.

General:

1. Any dispute or claim arising out of or in connection with this License Agreement or the Work (including non-contractual disputes or claims) will be governed by and construed in accordance with the laws of the U.A.E. as applied in the Emirate of Dubai. Each party agrees that the courts of the Emirate of Dubai shall have exclusive jurisdiction to settle any dispute or claim arising out of or in connection with this License Agreement or the Work (including non-contractual disputes or claims).
2. Your rights under this License Agreement will automatically terminate without notice and without the need for a court order if at any point you breach any terms of this License Agreement. In no event will any delay or failure by Bentham Science Publishers in enforcing your compliance with this License Agreement constitute a waiver of any of its rights.
3. You acknowledge that you have read this License Agreement, and agree to be bound by its terms and conditions. To the extent that any other terms and conditions presented on any website of Bentham Science Publishers conflict with, or are inconsistent with, the terms and conditions set out in this License Agreement, you acknowledge that the terms and conditions set out in this License Agreement shall prevail.

Bentham Science Publishers Ltd.
Executive Suite Y - 2
PO Box 7917, Saif Zone
Sharjah, U.A.E.
Email: subscriptions@benthamscience.org

BENTHAM SCIENCE

CONTENTS

FOREWORD

As it is well established, the second law of thermodynamics is a cornerstone of natural sciences. However, its impressive consequences cannot be considered as a closed issue. Despite having been established more than one hundred years ago, it still surprises us with its extensive reach and profound results, many times associated with simple and elegant arguments. Indeed, it is probable that its actual form will not change in its essential meaning even within the context of new possible scientific paradigms that will emerge during the development of science. This is the reason why it has fascinated the greatest minds for over a century. Its reach spreads all over the sciences: in the diffusion of an ink drop in a glass of water, in the chemical reactions inside of living things, attached to evolutionary processes; it may even be connected with the answer to the question: why do we live in a 3-dimensional space? The reason behind the wide reach of thermodynamics is that almost all phenomena around us can be summarized as how energy is transformed. Nature is strongly related to equilibrium and optimization, and despite the particularities of every field of knowledge, these two ingredients, in one way or another, will remain present. In this sense, it is a relevant task to bring thermodynamics and specially the concept of entropy closer to all the disciplines of science. The actual technological and industrial needs, everyday more demanding, require the compromise of the scientific community to work together, within interdisciplinary efforts, to understand and look for the better way to deal with the environmental impact and the best energy management in the new technologies, as well as establishing linkages between theoretical models and the real world.

I applaud the diversity of approaches in which knowledge is transmitted, especially those, such as this work, that look for the integration of people from different fields and levels of expertise in the subject. In this text, you will find a broad discussion of the second law of thermodynamics and entropy, focusing on the concepts and their understanding, but without losing formality. I personally enjoyed the actual and fresh perspective of this text and I'm sure that, as a reader of this book, you will find out strong elements to deepen your understanding of entropy.

Dr. Julian Gonzalez-Ayala
Universidad de Salamanca
Salamanca, 2017

Preface

The second law of thermodynamics is one of the most fundamental laws that govern our universe and is relevant to every scientific field studying the physical world. Nonetheless, the second law's application makes constant reference to entropy, one of the most difficult concepts to work with, and this is the reason why they are discussed almost exclusively in highly specialized literature.

Thermodynamic entropy has been rigorously examined by classical, statistical, and quantum mechanics, which provide several mathematical expressions for calculating it under diverse theoretical conditions. However, the concept of entropy is still difficult to grasp for students and even more for educated laymen. As a scientist in plant biology, I fall into the second category with regards to this subject. Indeed, I first wrote this introductory book for myself; to approach my work with greater awareness about its physicochemical implications, I felt I needed better insight into the thermodynamic considerations that underpin spontaneous processes and allow plants, as well as humans, to achieve and improve the capability to exploit the environment to their benefit. When I consulted the literature on this topic, I found that although there are very many papers and books on the subject, the thorough explanation that I was looking for was scattered throughout them. I then began taking notes, and when I was finally satisfied I realized that, once organized and suitably presented, they were potentially interesting for other people looking for detailed, but not too advanced, clarifications on entropy and the second law of thermodynamics. I believe that a better understanding of these concepts requires a more satisfactory verbal explanation than is generally provided, since, in my opinion, a verbal approach is the one closer to the understanding capability of students and non-experts. This is why this book is focused on providing a verbal account of entropy and the second law of thermodynamics. In this sense, I deem that, beside to the basic mathematical formulations, a consistent explanation in verbal terms can be very useful for the comprehension of the subject by people who do not have a full understanding of it yet. Thus, I eventually came out with the present work, targeted to students and non-experts who are specifically interested into this matter and have a basic knowledge of mathematics and chemistry.

With this book I attempt to offer an account of thermodynamic entropy wherein verbal presentation is always a priority. Basic formal expressions are utilized to maintain a rigorous scientific approach to the matter, though I have always tried to explain their meaning. The essential outlines for a verbal account of thermodynamic entropy are summarized in the last chapter. Such an outline is how I wish I had been taught the core concepts of this matter when I was first introduced to it. Therefore, I hope it can be of help for a general introduction to the second law of thermodynamics and the basic concept of entropy. The main text of the present work aims to demonstrate the validity of the proposed verbal presentation from a rigorous, scientific point of view, but it also represents a resource for insights on specific topics since the verbal approach is adopted throughout the text. Several examples illustrate the concept of entropy in its different expressions. A number of notes provide further clarification or insight into the content in the main text and the reader may skip them on a first reading.

With regard to the contents of the this work, I have highlighted that the best way to conceive thermodynamic entropy that I found in the literature was that of a function of "energy spreading and sharing" as suggested by physicist Harvey S. Leff. Herein I try to take this line of thought further to verbally unravel the concept of thermodynamic entropy and to provide a better verbal account of it. I propose that a useful definition of entropy is "a function of the system equilibration, stability and inertness", and that the tendency to an overall increase of entropy set forth by the second law of thermodynamics should be meant as "the tendency to the most probable state", that is, to a macroscopic state whose distribution of matter and energy is maximally probable (according to the probabilistic distributions of matter and energy, and also considering the eventual presence of constraints). Thus, with time, an isolated system settles into the most equilibrated, stable, and inert condition that is actually accessible. I have provided a wide overview to introduce the rationale for these definitions and to show that they are consistent throughout the various levels and applications of the concept of entropy. The key idea is to extract from known formal expressions of entropy the essential verbal outlines of this concept and to use them to elaborate a verbal presentation of entropy that can be of general utility to non-experts, as well as to educators.

CONFLICT OF INTEREST

The authors confirm that they have no conflict of interest to declare for this publication.

ACKNOWLEDGEMENTS

Declared none.

An Account Of Thermodynamic Entropy

An Account Of Thermodynamic Entropy

2

CHAPTER 1

Introduction

Abstract: Basic concepts are defined, such as what thermodynamics aims to, what a system is, which are the state functions that characterize it, what a process is.

Keywords: Adiabatic system, Boundaries, Classical thermodynamics, Classical mechanics, Closed system, Exchange of energy and matter, Heat transfer, Interactions, Isolated system, Macroscopic systems, Microscopic structure, Open system, Parcel, Processes, Quantum mechanics, Quantization of energy, State functions, Statistical mechanics, Surroundings, Thermal reservoir, Universe, Work.

Thermodynamics deals with the overall properties of macroscopic systems as defined by state functions (that is, physical properties that define the state of a body) such as: internal energy, E; temperature, T; volume, V; pressure, P; and number of particles[1], N. In addition to these properties, which are easy to understand, macroscopic systems are also characterized by specific values of entropy, a further state function that is more difficult to comprehend. Entropy can be calculated in many diverse theoretical conditions by several mathematical expressions; however, the concept of entropy is still difficult to grasp for most non-experts. A troublesome aspect of entropy is that its expression appears to be very different depending upon the field of science: in classical thermodynamics, the field where it was first defined, the conceptualization of entropy is focused on heat transfer; in classical mechanics, where many of the first studies were performed, entropy appears to be linked to the capability of an engine to produce work; its nature was then more precisely explained by statistical mechanics, which deals with the microscopic structure of the thermodynamic systems and studies how their particles affect the macroscopic properties of these systems; finally, quantum mechanics, by focusing on the quantization of energy and particles states, showed that the probabilistic nature of entropy, already highlighted by statistical mechanics, is closely dependent on the non-continuous nature of the universe itself. This works aims to show that, eventually, all these different aspects of entropy, which are necessarily linked to each other, can be better understood by considering the probabilistic nature of entropy and how it affects the properties of macroscopic systems, as well as the processes by which they interact.

A first exigency is then to define a macroscopic system. According to Battino *et al.* [2] a system is "any region of matter that we wish to discuss and investigate". The surroundings consist of "all other matter in the universe that can have an effect on or interact with the system". Thus, "the universe (in thermodynamics) consists of the system plus its surroundings". The same authors notice that "there is always a boundary between the system and its surroundings and interactions between the two occur across this boundary, which may be real or hypothetical" [2]. In any case, a thermodynamic system, which typically includes some matter, is a part of the universe that is clearly defined and distinguishable from all the other parts of the universe. The presence of solid boundaries around the system is an obvious aid in the identification of the system, but it is not a necessary one. A gas inside a vessel is traditionally seen as a good, simple thermodynamic system. An aquarium can be another quite well defined thermodynamic system, although the presence of fish and other organisms would greatly complicate its thermodynamic properties and entropy in particular. The Earth, even though it is not surrounded by solid boundaries, represents a thermodynamic system too since it is something that is clearly defined and distinguishable by the remaining universe: beyond its stratosphere there is extended empty space that separates our planet from other systems. Although some energy and matter can move through the theoretical boundary that divides the Earth from empty space, these can be precisely identified as transfers of energy and matter from/to the Earth system, which maintains its identity. It can then be immediately noted that the identification of a system is essentially a theoretical step, since the Earth can contain innumerable smaller systems, *e.g.*

vessels containing fluid, like an aquarium, but also each cell of a fish in an aquarium is clearly an enclosed system with physical boundaries.

Hence, a system must be clearly delimited to be studied, but the rules that govern a system must also hold for any macroscopic parcel of matter with corresponding properties and conditions. Simply put, a system is just an identifiable parcel of the universe. It is worthy to note, therefore, that any parcel inside a system has, with the rest of the system, the same relationship that holds between an open system and its surroundings. So, for a system to be equilibrated it is necessary that all of its parcels, however sorted out, are equilibrated with the rest of the system, just like a thermodynamic system equilibrates with its surroundings if there is no insulating barrier between them.

What is particularly relevant to the study of entropy is that the system that is under consideration has to be macroscopic, that is, it must include a huge number of particles. This is because of the above-mentioned probabilistic nature of entropy, which can really be appreciated when, in the presence of a large number of particles, the probabilities become determining for every feature of the system; that is, the intensive state functions of the system (*i.e.*, those that do not depend on the system size, but, rather, on the distributions of particles and energy across the system), which emerge as overall properties from the statistical average behaviours of all the constituting particles, can be sharply defined and statistical fluctuations due to random effects of the particles are so small that they can be neglected. Of course, the concept of entropy holds true for every system, including very small ones consisting of only a few particles. However, some statistical approximations and mathematical simplifications that are ordinarily used for large systems, cannot be applied when dealing with very small ones. Notably, in every system, intensive state functions like temperature, density and pressure, which are widely used to characterize large systems, can undergo instantaneous fluctuations. However, whereas such fluctuations are negligible in large systems, as we will see, they become relevant for very small ones. In fact, in very small systems, intensive state functions can attain ample inhomogeneities across the system itself, or between the system and its external environment if they are connected, even at equilibrium. Hence, very small systems cannot be accurately described by single average values for these parameters, which, thus, lose their physical meaning. The exact probabilistic distributions of particles and energy have to be considered for these systems, whose physical description requires, therefore, mathematical formulations that are much more complicate and less intuitive. So, though the verbal account of the entropy that will be provided in the last chapter is always valid, the formulations that must be used for precise quantification of entropy changes are more intuitive when large systems, like the ones we commonly are used to deal with in everyday life, are considered.

In addition, for the sake of simplicity, systems are also assumed to be at temperatures much higher than absolute zero, since some quantum effects that complicate the characterization of a macroscopic system arise at low temperatures, where the nature of the particles becomes more similar to that of a packet of waves than to an ideal, spherical, and dense unit of matter (actually, matter particles remain wavepackets even at high temperatures, but then they can be treated as if they were ideal, spherical, and dense units of matter).

Thermodynamic systems can be classified according to their capability to interact with the outside. An isolated system neither transfers energy or matter with the rest of the universe nor it changes its volume; a closed system can transfer energy to/from the outside and/or change its volume; an adiabatic system can change its volume, but cannot exchange energy or matter with the outside environment; an open system can transfer both energy and matter with the rest of the universe and can change volume too.

When studying the interaction of a thermodynamic system with its outside (obviously not in the case of an isolated system), it is important that the state functions of the system are known, and even the outside environment has to be clearly defined, since any interaction can cause a different result depending upon the conditions of the environment surrounding the system. Therefore, the state functions of the surroundings, at least those involved in the studied interaction, have to be defined as well. In many cases, temperature, pressure, or chemical potential for a given interaction must be defined for change in the system, specifically in the entropy, to be computed. Thus, the surroundings often act as a thermal reservoir that assures that the temperature remains constant. Analogously, in some processes, the pressure or the chemical potential is held constant. In this way, the interactions between a system and its surroundings can be measured and any change in the state functions can be properly accounted for. These interactions are called processes and are studied by thermodynamics to understand how the states of a system and its surroundings interact, as well as how physical processes generally occur. Clearly, in everyday phenomena, the theoretical and experimentally rigorous conditions assumed by thermodynamics do not commonly occur. Nevertheless, as is usual in science, results of experiments conducted in rigorously controlled conditions help us understand what factors are involved in a given phenomenon and thereby provide guidance to comprehend all similar phenomena that happen in the physical universe. Theoretical analysis assists us in defining to which actual instances such a general inference can be extended. Hence, if we understand how thermodynamic properties, and specifically entropy, affect systems and processes that are exactly defined from a theoretical point of view, we can then try to extend these findings to correctly interpret all the systems and processes of the universe.

CHAPTER 2

Entropy in Classical Thermodynamics: The Importance of Reversibility

Abstract: The concept of entropy is introduced within the context of classical thermodynamics where it was first developed to answer the questions: What happens when heat spontaneously goes from hot bodies to cold ones? To figure out this change, the transfer process must have some peculiar features that make it reversible. First of all, to be studied, systems and processes must be conceptualised as occurring at equilibrium conditions, or in conditions close to equilibrium. A first description of equilibrium is provided.

Keywords: Balancing of forces, Classical thermodynamics, Clausius' equation, Dynamic equilibrium, Entropy, Equilibrium conditions, Friction, Heat capacity, Heat dissipation, Ideal gas, Infinitesimal change, Irreversible process, Isothermal process, Potentially available energy, Quasi-static process, Reversible process, Stationary state, Thermal energy, Transfer of heat, Turbulence.

THE FOUNDING DEFINITION OF ENTROPY

In classical thermodynamics, entropy was originally defined by Clausius' equation [3]:

$$dS = \delta Q/T$$

where dS is an infinitesimal change in the entropy[2] (S) of a closed system at an absolute temperature T, consequent to a transfer of heat (δQ) that occurs at equilibrium conditions. This definition was prompted by the observation, among others, that heat is spontaneously transferred from hot to cold bodies and the direction and universality of this phenomenon could be explained by some change in the involved bodies. So, the transfer of heat has to result in a change in the

Alberto Gianinetti

bodies that necessarily drives heat transfer from hot to cold and not the reverse. Although theoretically correct, the adoption of heat transfer as a definition of entropy is troublesome for our intuitive comprehension, since it requires that the transfer has to be at a defined temperature, *i.e.* the temperature of the entire system has not to change, in other words, the transfer has to be isothermal. This is unfortunate because what typically happens when something is heated is just that its temperature increases. In fact, a transfer of heat is commonly expected to occur from a hot body to a cold one, but in the simple heat to temperature ratio pointed out by the Clausius' equation only one temperature is considered. Which one should be used, the temperature of the hot body or that of the cold body? As we are presently using the Clausius' equation to define the entropy of a system, if the temperature is diverse in the two bodies, or systems, their entropy changes are different as well, as we will see soon, and the entropy changes are separately calculated for each body, even though the amount of heat transferred from one system to the other is one and the same. However, what is relevant here is that the temperature of each body is changing. So, the question could be whether the initial temperature, the equilibrium temperature, or perhaps their mean should be used. Actually, when the process is not isothermal and therefore the overall transfer of heat is discrete rather than infinitesimal, infinitesimal transfers of heat can nevertheless be imagined to occur at each subsequent instant, that is, instantaneous transfers of heat can be considered, which are infinitesimal as well. Although this approach is theoretically correct, and it will be used later in this chapter to show what generically happens to the total entropy when the transfer is not isothermal, in this case the problem is that the temperature at which each heat transfer occurs is changing for the system itself and is therefore unknown, unless further assumptions can be done for modelling it. As these assumptions depend on the specific bodies, it is not desirable to consider them when dealing with the general principle of equivalence between heat transfer and entropy change. In addition, if the initial temperatures of the bodies are diverse, a gradient of thermal energy forms during the equilibration process (unless it is extremely or, better, infinitely slow, as we'll see), thus generating inhomogeneities across the system itself, a problem that has already been remarked when discussing the size of the system and that greatly complicates the analysis of instantaneous changes. So, the problem with a non isothermal heat transfer is that the value of the Clausius' equation is different for the two bodies, and it is indeterminate inasmuch the instantaneous temperatures of the bodies are unknown, or indefinable. Obviously, for each body the equation holds true at every instant, it is just a problem of not being able to calculate it for a generic body if its temperature changes during the heat transfer. This is bothering, since the differential Clausius' equation just cannot be experimentally tested, or directly applied, in this situation. A theoretical solution to this methodological difficulty is based on the fact that the amount of transferred heat becomes smaller and smaller as the temperatures of the two bodies get closer. In the limit of the difference between the two temperatures becoming infinitesimal, even the amount of transferred heat becomes infinitesimal. If the difference of temperature that drives the transfer of heat is infinitesimal, the process is actually isothermal. Thus, the Clausius' equation is always true, but it provides a definable and exact measure of the change of entropy of a body, or system, when it is generated at a given, unchanging, temperature of the system, and this condition is unfailingly guaranteed by considering an infinitesimal transfer of heat.

A discrete version of the Clausius' equation is even more problematic: for a discrete heat transfer, the T in the denominator of $dS = \delta Q/T$ is not constant and, therefore, not only dS cannot be calculated using this simple ratio, but the Clausius' equation itself does not hold true unless the process is isothermal. The reason for the isothermal requirement is that entropy is a non-linear function of temperature (it is logarithmic, for an ideal gas, as can be found out by integrating the Clausius' equation) and therefore the actual instantaneous change of entropy is different at different temperatures[3]. Thus, since in many processes it is not possible for the temperature and thermal energy of an ideal gas, or other system, to change independently of one another, a change in the entropy of a system can be expressed properly in terms of the simple heat to temperature ratio only in differential terms, specifically, $dS = \delta Q/T$ (and not $\Delta S = \Delta Q/T$, unless, as we are going to see, there is a way to buffer any change of T) or its equivalent [4].

HEAT CAPACITY IS REQUIRED TO CALCULATE THE ENTROPY CHANGE IN DISCRETE HEAT TRANSFERS

The classical kinetic theory of gases implies that the amount of thermal energy possessed by a mole of an ideal gas (*i.e.*, a gaseous system) would be directly proportional to the system temperature, specifically, such a proportionality depends on c_v, the constant volume heat capacity [4]. The heat capacity can be seen as the capacity of a body, or system, to accept heat, actually thermal energy, for its temperature to increase of one degree. So, a body with a large heat capacity needs a lot of heat, *i.e.* thermal energy, to increase its temperature. More formally, the heat capacity is the coefficient of proportionality between the temperature of a body and its internal energy when only thermal interactions are involved. Since, when the temperature increases, an ideal gas increases either its volume or its pressure, it is usually necessary to specify which one of these two state functions is allowed to change in the studied system, to properly characterize the system itself. The constant volume heat capacity, c_v, relates changes in temperature to changes in internal energy. Thus,

$$dE = c_v \cdot dT$$

and c_v would be a constant. This equation can be expressed in the form;

$$dS = c_v \cdot dT/T = c_v \cdot d\ln T$$

and integrated between two specific temperatures to yield;

$$\Delta S = c_v \cdot \ln(T_2/T_1)$$

The same difficulties occur for a change of entropy consequent to a finite change in the system's volume (and then in the concentration of the particles), which is given by the comparable expression;

$$\Delta S = R \cdot \ln(V_2/V_1)$$

which is obtained essentially in the same manner [4].

So, as previously discussed, it is not possible to directly use the Clausius' equation to calculate a discrete change of entropy if the temperature is changing because S does not change linearly to T. Discrete changes of entropy in a system, consequent to a discrete change of either temperature or volume, can instead be calculated from the logarithmic equations reported above, assuming the respective proportionality constants remain really unchanged. However, this approach introduces an additional parameter, precisely the proportionality constant, that, essentially in the case of the logarithmic equation for temperature change derived from the Clausius' equation, is dependent on the specific system, or body. As said, this is undesirable for a general definition of entropy, which in the simple ratio established by the Clausius' equation has one of its most general expressions and its founding definition. This difficulty (*i.e.*, the isothermal requirement for the application of the Clausius' equation) can be overcome in three ways: either the transfer of heat has to be infinitesimal, as seen, so that any change in the temperature is negligible (in other words, the entropy change at each temperature is calculated as the limit of entropy change when heat transfer tends to zero, *i.e.* as the local derivative of entropy *vs.* temperature and this is a more general form reflected in Clausius' equation), or any change in the temperature consequent to heat transfer from the system to the surroundings is buffered because the latter act as a very large (theoretically infinite) thermal reservoir [4], or there must be some other reversible transformation (we will see below what is specifically meant for a transformation to be reversible) that maintains a constant temperature even if heat is added. In the latter case, some suitable heat-consuming process, like an isothermal expansion or a hypothetical phase transition that may use the heat to melt a solid phase without increasing the temperature, and without any volume change (the use of an ice/water system at 0°C is therefore only roughly suitable because when freezing water expands its volume; this is why ice is less dense than water and floats on the surface of water), can be coupled to the heat transfer to avoid a discrete heat transfer causing a change of the temperature of the system.

In actual experiments, if the transfer of heat at each given temperature is (theoretically) infinitesimal, then the cumulative change of entropy occurring in the system as a consequence of a discrete, non-isothermal heat transfer (which would then require an infinitive time to be accomplished), can still be calculated from the integration of the heat transfer over the continuous change of temperature. However, as already mentioned, a fast heat transfer is problematic to deal with, because of the inhomogeneity of the thermal energy in the system that makes the system temperature indefinable. So, when dealing with real conditions, a very slow change, ideally occurring through a series of equilibrium states, is required for the heat transfer and then the entropy to be properly measured. That is, the process has to be reversible (shortly, this means it can go either direction). In fact, if a system is not in equilibrium it is not even possible to define its temperature [1].

REVERSIBILITY

At this point it is evidently necessary to clarify when a process is reversible and, subsequently, we will see why it needs to be so. To be reversible a process must be such that reversing the path at every stage of the process restores both the system and its surroundings to their initial conditions (as defined by the state functions)[5]. Typically, this occurs when a process is at equilibrium, meaning it can go one or the other way, indifferently. This means that any transfer between the system and its surroundings does not have a spontaneous or preferred direction, but it is a fluctuation of a dynamic equilibrium. In addition, a process at equilibrium is expected to stay at equilibrium and to go nowhere else. Nevertheless, a reversible process is intended as an ideal, discrete transfer that occurs through a series of equilibrium states and actually goes toward a given direction. How can a series of equilibrium states go toward a given direction? It can only happen if any change occurring in the system toward that direction does not affect the equilibrium itself. For example, if in a system at the phase transition of water (0 °C) there are two ice cubes, one small and the other big, the equilibrium is not affected if, because of random fluctuations of the dynamic equilibrium between the molecules of the ice and the water, very gradually the previously small ice cube becomes larger and the prior bigger one becomes smaller. It would be a transfer the equilibrium is indifferent to. In general, a transfer of energy or particles from/to a system, as well as from one form to another perfectly equivalent form, does not affect the overall equilibrium, even if it is macroscopic. Thus, a process is reversible if it causes a change from a given condition to another perfectly equipollent in energetic terms. As we will see, this means that there is no overall change of potentially available energy, even considering the possible inter-conversions between diverse forms of potential energy (gravitational, hydrostatic, PV, thermal, electrical, *etc.*). So, the essence of a reversible process is not that it stays at a given fixed state, but rather, that all forces are balanced, *i.e.* it is at equilibrium at each and every infinitesimal step [2, 3]. The series of equilibrium states that lead to change, or transfer, should then be theoretically infinite in mathematical terms and very slow in actual terms. What is fundamental is that a reversible process is intended to occur, but it cannot be guaranteed since any admissible change must not affect the equilibrium. Change could happen, but it could also not proceed toward the intended direction. If a process is to occur in a desired direction, then there must be an imbalance of forces, however small, to push it in that direction. Therefore, reversible processes are discussed from a purely theoretical point of view. Nonetheless, they can be approximated in practice.

In practice, a reversible process can be approximated if some conditions are satisfied. Before anything else, the process has to be very slow, so that the system and its surroundings always approach a stationary state. In other words, a reversible process can be approximated by a sequence of states wherein any intensive state function (temperature, pressure, chemical potential) of the surroundings changes slowly enough to allow the system to continually adjust and equilibrate after each tiny change. This is often defined as a quasi-static process, and it goes in a given direction through a finite, or practically continuous, number of equilibrium states [2]. Actually, this ensures that both the process goes in a desired direction, and the system is almost in an equilibrium state with its surroundings throughout the process; however, it pays no attention to how the overall process is generated. Typically, the cause of the process occurring in the surroundings is ignored because the thought-experiment is aimed to evaluate the irreversible path for the system by itself, as a paradigm for irreversible processes in general. Causation is therefore abstracted away to avoid going back along an endless cascade of causal events.

A quasi-static process is formally irreversible because all forces are not balanced at each and every finite step, since this is the only way to assure that the process ultimately proceeds in a given direction. Directionality, caused by this tiny unbalancing, is what distinguishes a quasi-static process from an ideal, reversible one. However, ultimately, the quasi-static finite stepwise process approaches the reversible one, because the process goes in one direction in a very slow manner so that at each instant it is almost balanced (that is, the system and its surroundings are almost balanced). A quasi-static process should virtually replicate in practice what a reversible process is in theory, at least as regards the system. In addition, if a system is not at equilibrium it is neither possible to exactly define its temperature nor to consider its thermal properties as homogeneous. Inhomogeneities in temperature obviously have a complex effect on any thermal process. In general, the value of an unknown state function can be computed only if the system is equilibrated (and homogeneous for that state function) and therefore any ongoing process can only be properly quantified if it is reversible.

Thus, to precisely quantify the change of entropy that occurs in a heat transfer between a system and its surroundings (typically considered to act as a thermal reservoir ensuring a constant temperature T), theoretical considerations require that the transfer occurs at equilibrium conditions; a discrete change in entropy could be obtained through an infinite series of incremental steps, δQ, so that at each instant the system is at equilibrium and therefore:

$$\Delta S_{sys} = \int \delta Q / T$$

providing that the reversible heat transfer and the corresponding change of temperature are continuously monitored, the discrete entropy change can be calculated as a sum (integration) of (theoretically infinitesimal) steps. Notably, a series of equilibrium states occur only if the temperatures of the system and its surroundings are always equilibrated. In fact, heat flows from hot to cold and the flow rate goes to zero at equal temperatures [1]. In practice, this ideal reversible process can be approximated if the heat transfer, as well as any subsequent transformation aimed to keep the temperature constant, is very slow so that the system and its surroundings always are in an almost stationary state.

Entropy is an extensive state function, *i.e.* it depends on the spatial extension of the system and on its overall actual state, and therefore it does not depend on the pathway followed to reach a given state whether the process is reversible or irreversible. However, irreversible processes increase the combined entropy of the system and its surroundings: a process that involves friction, or that is done quickly and generates turbulence in fluids, is irreversible also because some heat is dissipated into the surroundings thus that there is an overall increase of entropy (for example, when a falling body impacts the soil). Anyway, a change of entropy in a system that undergoes heat transfer is the same whether the transfer of heat is reversible or not, providing the end temperature is unchanged. So, reversibility does not affect the change in entropy of a system if its starting and final conditions are the same, but if the transfer is irreversible, that is, it occurs in non-equilibrium conditions, then there is a discrete difference in the temperature between the system and its surroundings. Thus, there is a direct transfer between a hot body and a cold body and the resulting changes of entropy in the two bodies (or system/surroundings) do not correspond. In fact, if the heat Q flows directly from a hot body at T_H to a cold one at T_C, the respective gain and loss of entropy for the two bodies are $\delta Q/T_C$ and $-\delta Q/T_H$, thus that, as $T_H > T_C$, according to Clausius' equation the decrease of entropy in the hot body is lower, as an absolute value, than the increase of entropy in the cold one. Thus, in a heat transfer, the changes of entropy in the two bodies compensate for each other only if the process is reversible and isothermal, otherwise, the increase of entropy in the cold body is higher than the decrease in the hot one and the overall entropy then increases. For a direct transfer of heat from hot to cold the total variation in entropy is:

$$dS_{tot} = \delta Q/T_C - \delta Q/T_H$$

which is always positive [6]. Hence, only a reversible path provides a valid equivalency between the measured heat transfer and the entropy change in both a system and its surroundings. In general, for every kind of process and not only for heat transfers, there is always an increment of the overall entropy in going from a nonequilibrium to an equilibrium state. As a result, processes that are not reversible increase entropy. In a discrete but idealized and reversible transfer of heat to a system, $\Delta S_{sys} = \Delta Q_{rev}/T$, but in a real, direct heat transfer to a cooler body $\Delta S_{sys} > \Delta Q_{irrev}/T$, where T is the end temperature of equilibration (for example, with a thermal bath), because the temperature of equilibration must be higher than the one the system initially had for the transfer of heat to the system to be spontaneous and discrete. This is another way to show that the discrete version of the Clausius' equation does not hold true if T changes.

Definitively, for a process to be reversible means that it occurs under equilibrium conditions, that is, it involves some change or transformation that does not affect the overall equilibrium of the system. Otherwise, some additional change happens in the system when it moves to equilibrium and such a change alters the balance of the overall entropy. In addition, if and only if the system is at equilibrium are its state functions defined and homogenous throughout the system. This allows us to obtain exact measurements of the state functions and then to establish precise relationships between them.

Classical thermodynamics looked for a state function that could describe what happens when heat spontaneously goes from hot bodies to cold ones. It was soon found that an increase of entropy is always associated with this process.

CHAPTER 3

Heat *vs.* Work

Abstract: The idea of transferring energy from/to a system is expanded to include work in addition to heat. Work is an alternative form by which the energy of a system can be transferred.

Keywords: Boltzmann constant, Classical mechanics, Conservation of energy, Dissipation of energy, Entropy increase, First law, Heat, Intermolecular potential energy, Internal energy, Kinetic energy, Microscopic level, Thermal energy, Transfer of work, Waste heat, Work.

Whereas in classical thermodynamics the interest was focused on heat transfers, in classical mechanics the attention was shifted toward the potentially much more applicative theme of the relationship between heat and work, with particular regard to how thermal energy can be used to produce work.

It is useful, at this point, to examine in a more in-depth manner the concepts of heat and work as thermodynamic ways of transferring energy. The first law of thermodynamics is often formulated by stating that any change in the internal energy (*i.e.*, the energy contained within a system, excluding the kinetic energy of motion of the whole system and the potential energy of the system as a whole due to external force fields, and, with the exception of nuclear reactions even the rest energy of the matter, $E = mc^2$, is not considered because in common systems there is a clear distinction between matter and energy [7]) for a closed system is equal to the amount of heat supplied to the system, minus the amount of work done by the system on its surroundings. In other words, the energy is additive and an exact balance can be calculated when it is exchanged. Thus, the first law of thermodynamics is essentially the statement of the principle of the conservation of energy for thermodynamic systems [6, 8]. The sum of the heat (Q) and work (W) associated with a process represents a change in the internal energy, which is a state function of a system, as defined by the First Law of thermodynamics:

$$\Delta E = Q - W$$

where Q is the heat received by the system and W is the work done by the system. This is equivalent to the expression: $\Delta E = Q + W$, wherein the work done on the system is considered (this is just a matter of definition). Thus, if an ideal gas expands reversibly (*i.e.* against an external pressure, to equilibrate the system pressure) and isothermally (*i.e.* in the presence of a thermal reservoir), work is done by the gas and such work is equal to the heat absorbed by the gas during expansion (Fig. **1A**). Hence, a certain quantity of heat is converted entirely into work. At the same time, the entropy of the system has increased by some amount while the entropy of the heat reservoir has decreased by the same amount (since the heat transfer is isothermal and the work expended by the system is the same received by the surroundings). Thus, this symmetric change of entropy is equivalent to the transfers of heat and of work. However, real processes are irreversible and a macroscopic feature of these processes is that they are associated with an overall entropy increase (Fig. **1B**). Indeed, any transfer of work in real systems always implies that some heat is also produced as by-product, that is, as waste heat. Specifically, when a process exploits a difference in temperature any transfer will be of heat or of heat plus some work (which reaches a maximum proportion for reversible processes). Any transfer of heat (through a temperature gradient) that does not occur at equilibrium conditions, *i.e.* that does not produce an equivalent amount of work to build up another equipollent modification that maintains unchanged the overall amount of available energy, evidently results in a dissipation of available energy (*i.e.* of the energy potentially available from the temperature gradient to do work) and inevitably into an increase of entropy.

A transfer of heat is a transfer of thermal (or kinetic) energy [5]. The thermal energy is the sum of all the translational, rotational, and vibrational energies of the particles in a system.

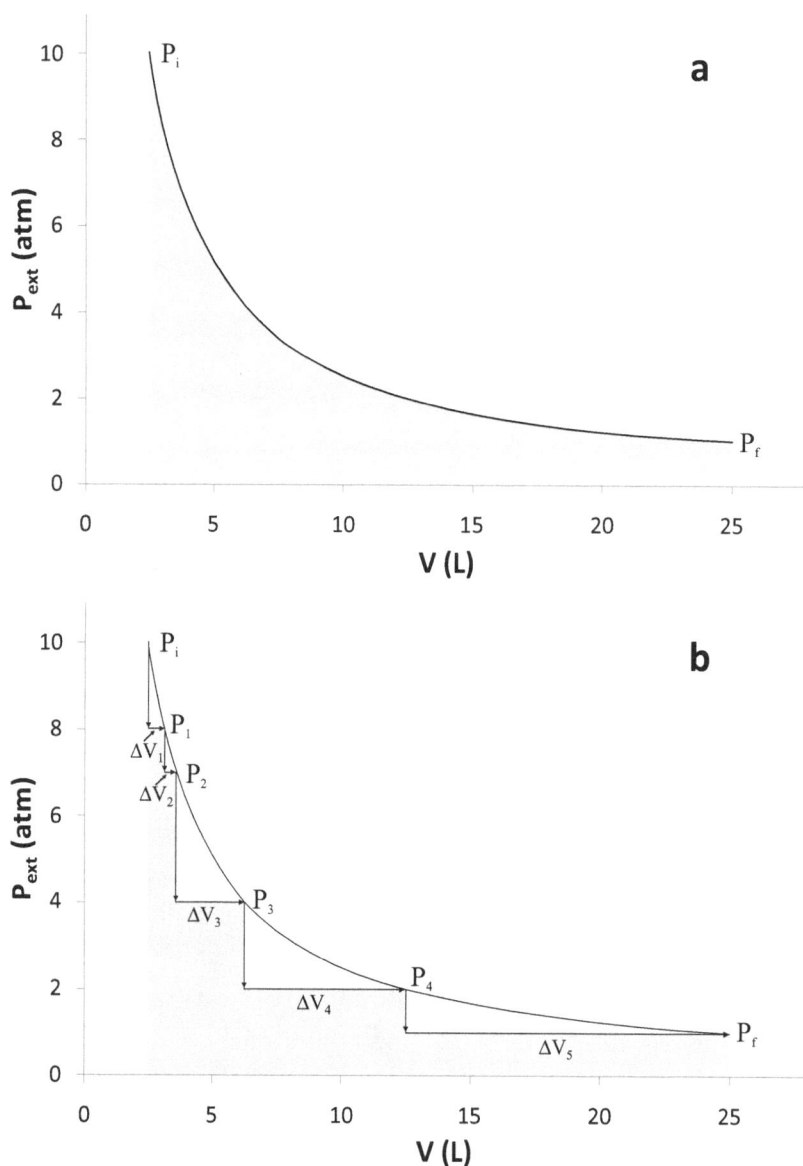

Fig. (1). An indicator diagram for the isothermal expansion of a closed system consisting of 1 mole of an ideal gas (at temperature $T = 304.66$ K). P_{ext}, the pressure exerted on the system by the surroundings (which also act as a constant-temperature thermal reservoir), is plotted against the volume V of the system [5]. For (**A**), in the reversible path, the gas is always pushing against the slowly decreasing external pressure, P_{ext}, which then remains equal to the pressure of the system. The PV work is done reversibly to the maximum possible level and is represented by the area under the curve, defined by the reversible isothermal equilibrium $PV=nRT$, between the initial P_iV_i and final P_fV_f conditions; it is given by $W_{rev} = nRT \cdot \ln(P_f/P_i) = -57.56$ L·atm. In (**B**) an irreversible path (stairstep plot) consisting of five steps of abrupt pressure drop, $-\Delta P_{ext}$, compared with the reversible path (thin smooth curve). For this irreversible process, temporary pressure gradients occur in the gaseous system. However, by using P_{ext} the work done on the surroundings can easily be computed [5, 10]. Each irreversible step has two stages: an isochoric P_{ext} drop, which is isothermal as no work is done, is immediately followed by the second step, an isobaric expansion of the gaseous system, with the gas pushing against the new fixed level of P_{ext} (each expansion is isobaric as far as P_{ext} is concerned, since the pressure applied by the surroundings is constant while the system expands [5]). During the expansion, the gas does work on the surroundings, which is compensated for by an equivalent heat transfer to the system, thus even the expansion is isothermal. The PV work done is represented by the area under the stairstep plot and is smaller for the irreversible path than for the reversible path because of the pressure lost in the P_{ext} drops. Correspondingly, the heat transfer for an irreversible isothermal expansion differs from the heat transfer for a reversible isothermal expansion between the same endpoints. In fact, as $|W_{irr}| < |W_{rev}|$ and $\Delta E = 0$ for the isothermal expansion of an ideal gas, we also have $|Q_{irr}| < |Q_{rev}|$ [5]. Specifically, the work produced by the irreversible path is given by the sum of the areas beneath the stairstep plot: $W_{irr} = -(P_1 \cdot \Delta V_1 + P_2 \cdot \Delta V_2 + P_3 \cdot \Delta V_3 + P_4 \cdot \Delta V_4 + P_f \cdot \Delta V_5) = $ -43.84 L·atm, with an efficiency of 76.2%. It is worth noting that, for each, step, the external pressure for each isobaric expansion is used in the computation, since it is the pressure against which the system does the work of expansion. Then, any drop in P_{ext} represents lost work and an overall entropy increase, since it corresponds to an irreversible free expansion, which does not produce work. Of course, the pressure drop is not that sharp in real gaseous systems, but rather, after an initial major fall, the system pressure approaches P_{ext} more gradually until the two pressures equilibrate. It is precisely this temporarily pressure difference that drives the expansion of the system. Nevertheless, no additional work is obtained in a real process by a system pressure exceeding P_{ext}, because any unbalanced expansion is as much free as it is unbalanced, and, in addition, any friction, as well as the turbulence consequent to each sudden pressure drop, generates a small amount of waste heat that dissipates into the surroundings and further decreases the work really available. Hence, in the irreversible path, work is lost because of both unbalanced expansion and waste heat. The degree of irreversibility is 100 - 76.2 = 23.8%, and the corresponding overall change of entropy is $\Delta S = W_{lost}/T = (|W_{rev}|-|W_{irr}|)/T = $ 1390.179/304.66 J/K = 4.56 J K^{-1} (where work has been expressed in Joules). Clearly, this does not include the entropy increase eventually associated with the process causing the pressure drops in the surroundings (after all, the drops have been caused in some way and they are evidently not quasi-static), but only includes the system's 23.8%-irreversible expansion that these drops have provoked in the system.

Heat is directly linked to the amount of thermal energy possessed by a given mass of matter, but since each substance needs a specific amount of heat for each mass unit to increase its temperature by one degree under the same conditions, it turns out that different substances, which are by definition composed of different molecules (which show varying kinds and degrees of inter-molecular interaction forces), have diverse heating requirements to reach the same temperature (because these interaction forces tend to reduce the freedom of movement of the particles and therefore have to be overcome with some expenditure of thermal energy for it to display in terms of translational, rotational and vibrational energies). Therefore, $dT = \delta Q/c_v$, where c_v is the constant volume thermal capacity of a given matter, body, or system.

Thus, kinetic (or thermal) energy is the energy associated with the motion (translational, vibrational, and rotational) of a molecule. For particles with radial symmetry, like atoms, rotational motion can be ignored. In general, a system is defined by a given internal energy, E, which, at the microscopic level, includes kinetic energy due to the microscopic motion of the system's particles and the intermolecular potential energy (stored energy) associated with the microscopic forces between the particles, including the energy stored in the chemical bonds and in the physical force fields of a system, due to internal induced electric or magnetic dipole moments, *etc.* (and, if nuclear reactions are studied, also with the static rest mass energy of the constituents of matter, explicitly, the nuclear energy stored by the configuration of protons, neutrons, and other elementary particles in atomic nuclei). As the internal energy of the system is conserved, the sum of the kinetic and intermolecular potential energy must be constant. This means that every change in the kinetic energy of a system must be accompanied by an equal but opposite change in the intermolecular potential energy. In chemical processes this is typically in regards to the chemical potential energy of the intermolecular forces among the particles of the system [1]. This would seem to mirror the first law of classical thermodynamics that states that any change in the internal energy of a closed system can only be caused by a transfer of work or heat. However, the first law looks at the macroscopic behaviour of a system interacting with its surroundings, that is, when transfers of E are considered from/to the system, whereas the distinction of E into kinetic and intermolecular potential energy considers the microscopic situation of the system. Clearly, since within both the two decompositions of E (heat + work and kinetic + potential) there is interchangeability between the components, there is also no defined direct relation between the constitution of E at the microscopic level and the transfers of E at the macroscopic level.

At the microscopic level, the kinetic component of the internal energy is dependent on the temperature of the system; this component is called kinetic energy, or thermal energy, and whose level is given as $k_B T$, *i.e.* the product of the Boltzmann constant k_B (it equals to the universal gas constant[5], R, expressed per particle instead of per mole: $k_B = R/N_A = 1.38065 \cdot 10^{-23}$ J K^{-1}, where $N_A =$

Avogadro's Number $= 6.022 \cdot 10^{23}$ mol^{-1}) multiplied by the absolute temperature T. The first law of thermodynamics states that, at the macroscopic level, the increase in internal energy of a system is equal to the total heat added and work done on the system. Therefore, if the system is isolated its internal energy cannot change. At $T = 0$, the internal energy is E_0, the ground state energy, which in chemical systems is frequently assumed to be zero, at least for an ideal solid, so that at higher temperatures, for an ideal gas (wherein there are no intermolecular forces among the particles, and then no intermolecular potential energy):

$$E = E_0 + cT$$

where c is a proportionality coefficient, represented by the thermal capacity of the system. Specifically, for an ideal gas at constant volume:

$$E = E_0 + c_v T$$

where $c_v = 3/2Nk_B$ is the constant volume heat capacity [4, 9]. If it is assumed that $E_0 = 0$, then the expression is further simplified to:

$$E \propto k_B T$$

The presence of intermolecular forces among the particles of the system affects the increase of T when heat is provided, which means that although E is dependent on the temperature because its kinetic component is proportional to T, the intermolecular potential energy affects this proportionality. The thermal capacity of a given, real system, c_v, includes this proportionality, at least for small changes of T, wherein c_v itself does not change with temperature.

Shortly, classical thermodynamics, partially converging with classical mechanics, aimed to answer two questions: why does heat goes from hot to cold, and what happens to available energy, potentially capable of producing work, when it goes lost as waste heat and dispersed to the environment without increasing the temperature? An increase of entropy is the answer to both questions, which appear as different facets of the same problem.

CHAPTER 4

More on Reversibility

Abstract: Once the concept of work has been introduced, the concepts of reversibility, equilibrium, and entropy become clearer and can be better defined. This was the task undertaken by classical mechanics.

Keywords: Quasi-static process, Reversibility, Frictionless process, Local equilibrium, Maximum work, Overall increase of entropy, Irreversible process, Cyclical process, Carnot cycle, Maximum theoretical efficiency, Work accumulation, Engines, Real process, Particle motions, Balancing of forces, Energy levelling, Maximum entropy, Equilibrium, Stable state.

Now that the nature of work has been introduced, it can be noticed that reversible processes can also occur without any transfer of heat. In fact, for an ideal gas, an adiabatic reversible expansion occurs if the system is thermally insulated but its volume changes very slowly against an external pressure, so that there is always a quasi-static equilibrium between the pressure of the system and its surrounding environment (which restrains the system's expansion). In this way, the pressure and temperature of the system decrease while its volume increases. The temperature of the system decreases because reversible work is done by the expanding system on the surroundings (*e.g.*, some kinetic energy of the system's particles is transferred to the expanding wall to move it against the kinetic energy of the surroundings' particles), that is, in the system heat is converted into work (on the contrary, in an irreversible free adiabatic expansion the temperature of an ideal gaseous system remains constant because no work is done[6]). If the quasi-static process is frictionless, the gain in entropy due to the volume increase of the system is exactly balanced by the loss of entropy due to the decrease of temperature, as no exchange of heat occurs with the outside. Thus, the difference in the system's entropy at two temperatures can be measured in terms of volume change. The change in entropy of the surroundings is considered to be zero with the assumption that the slow drop in pressure is caused by some reversible process. Hence, no overall entropy change occurs. Whereas, in the irreversible free adiabatic expansion of an ideal gaseous system into vacuum, the entropy of the system increases with the increase in volume, as no work is done on the surroundings and therefore the system temperature does not decreases. No work is done because the gas is not pushing anything out of its way.

In general, for a process to be reversible it necessitates that [1]: (i) if heat is transferred, the temperatures must be equal; (ii) if volume is transferred, the pressures must be equal; (iii) if particles are transferred, the chemical potentials must be equal. In practice, this means that at each point in time the system is infinitely close to a local equilibrium [9]. If any of these conditions are missing, some work is lost and the process results in an overall increase of entropy (Fig. **1B**). In addition, to be virtually reversible, quasi-static processes (specifically those that involve a volume change) must be frictionless, so that no additional energy is dissipated as waste heat. If a process is not frictionless, then some work done by the system is no longer available in the surrounding environment, which can be described as [10]:

$$W_{sur} = (W_{sys} - \text{friction})$$

and then;

$$|W_{sur}| < |W_{sys}|$$

In other words, we obtain the maximum work from processes that generates no entropy in the universe. Putting everything together, reversible processes are always the most efficient. They represent the theoretical limit on the maximum work that can be performed by systems [9].

An overall increase of entropy is therefore always associated with an irreversible process since it occurs under non-equilibrium conditions (and, typically, is not frictionless). As originally noted by Clausius [3], an irreversible transfer of heat generates an equivalent change in the bodies involved that always drives the heat transfer from hot to cold and not the reverse. In general, an irreversible process occurs toward a spontaneous direction and such a process is spontaneous only if it increases the overall entropy of the system and the surrounding environment. Under equilibrium conditions, any change of entropy in the system is equal in magnitude, but opposed in sign to the change in the entropy of the surroundings, so that there is no overall change in entropy. In fact, the second law of thermodynamics points out that any process that would reduce the entropy of an isolated system is impossible. A process that could go in either direction without entropy change is possible and is said to be reversible. However, in an irreversible heat transfer the overall entropy is increased, that is, some energy is dissipated in the moving of the process toward its spontaneous direction.

For a cyclical process, which is pertinent to engines as studied by classical mechanics, we have:

$$\Delta S_{tot} = \oint \delta Q/T \geq 0$$

that is, for an engine that undergoes a cyclical process, the sum of all instantaneous transfers of heat from the state at one point of the cycle to the same state at whatever subsequent cycle (this is indicated by the cyclic integral, $\oint$) is null (in fact, if the state is the same, the internal energy of the system, which is a state function, does not change, by definition, even though the internal energy changes during the course of the cyclic process), and therefore the entropy (which is a state function as well) of a cyclical engine (which is considered a system) does not change from one cycle to another (of course, considering the same point of the cycle). Thus, the net entropy change of a system subject to a cyclic process is always zero, whether the process is reversible or irreversible, as the state of a system is independent of which thermodynamic path was followed to reach it. However, whereas in a reversible cyclic process the entropy change of the surroundings is also zero (since the entropy change in the surroundings equals the change in the system and the surroundings themselves return to their initial state, or to one with the same entropy), in an irreversible cyclic process the entropy change of the surroundings (specifically, the thermal reservoirs) is always positive, and therefore the total change of entropy is positive too. In general, the change of entropy in the system plus its surrounding environment is positive in any real process and is zero only for ideal reversible processes, such as the reversible Carnot cycle, which is the most efficient heat engine cycle allowed by physical laws. In this cycle a net transfer of heat from a hot to a cold reservoir[7] produces utilizable work because the isothermal changes in volume (expansion and compression) that occur when the system is in contact with one or the other reservoir (hot and cold, respectively) can produce work (if, for example, a rotating wheel is properly connected) and the passage from one temperature to the other is mediated by corresponding adiabatic changes of volume, wherein the entropy does not change as work is transferred and the changes are compensated for by equivalent temperature changes[8]. Actually, as shown just by the Carnot cycle, even under reversible conditions, heat cannot fully be converted into work and the transfer of heat leaves some unused heat, thus that the maximum theoretical efficiency of a heat engine is described as:

$$\eta = 1 - T_c/T_h$$

where in, T_c is the temperature of the cold reservoir and T_h is the temperature of the hot reservoir. Only if the cold reservoir is at 0 K is the efficiency, in terms of conversion of thermal energy and for a given expansion-compression process[9], at 100%. To avoid any direct, spontaneous transfer of heat, the temperature of the cold reservoir must match the minimum temperature reached at the end of the adiabatic expansion, which depends on the maximum volume of expansion. The temperature and pressure of the surroundings, or cold reservoir, therefore place a limit on the maximum theoretical efficiency of a heat engine. Even though, in the Carnot cycle, the net heat flow from the hot reservoir to the cold reservoir represents an increase in entropy, the work output, if accumulated throughout the cycle, represents an equivalent decrease in entropy that could be used to operate the heat engine in reverse (*i.e.* as a cooling device) and return it to its previous state, thus the total entropy change is still zero at any instant in the cycle. There is, therefore, a negative equivalence between accumulated work and entropy. So, for an entire cyclical process (not just for a transfer of heat) to be virtually fully reversible there must be a third requirement, beside to being quasi-static and frictionless: the obtained work has to be accumulated and conserved in terms of usable energy (work by itself is a process and is therefore something not directly storable), otherwise it cannot reverse the process. For example, if work is used to lift a body in a gravitational field, a controlled quasi-static fall of this body can then be used to reverse the process (in the absence of any friction). If, on the other hand, the work is used to move (accelerate and decelerate) a locomotive through a flat prairie, at the end of the travel there is no way the process can be reversed without employing additional fuel (even if the travel was quasi-static and frictionless) since the work was done, but not accumulated. It is, however, in human nature to assume that, once some work has been obtained in a manner that could be reversible, one does not have to worry about whether the use of that work is really reversible or not. Hence, the third requirement is typically ignored and a process is commonly defined as reversible if the work it provides might, at least hypothetically, be used reversibly. This is why when considering a specific reversible process occurring, for example in a closed system, it is very often implicitly understood that any other referred change is also occurring at equilibrium conditions. For example, if the temperature, pressure, or another state function changes very slowly in the surroundings, a consequent quasi-static process is intended to occur in the system and, at the same time, even if this is not clarified, how the change of the state function actually occurs in the surroundings is neglected and/or tacitly assumed to happen in some reversible way (Fig. **1**).

However, in real engines, as well as in any real process, there is always some friction, turbulence, and/or leaking of heat that reduces the efficiency, commonly by a great deal, with respect to the theoretical maximum set forth by the ideal reversible process. As originally remarked by Kelvin, a transformation whose only final result is to transform into work heat extracted from a source, which is at the same temperature throughout, is impossible [6], thus a cyclic engine that only converts heat into work is infeasible.

Similar considerations can be adopted for electrical processes, among others: batteries are just devices that are aimed to accumulate energy, in the form of electrochemical potential (which is touched upon later), that is, they can be used to store mechanical work (actually, they store potential energy available to do work). The interconvertibility between different forms of potential energy offers many solutions to the problem of storing potential energy and therefore potential work. Accumulators, though typically not providing a reversible process, allow for the reduction of dissipation of energy and/or to make energy available when other sources are not at hand (*e.g.*, at the time of starting an internal-combustion engine).

We can now summarize this information about reversibility. At the microscopic level, thermal energy is linked to particle motions, that is, the greater the thermal energy the stronger the motion of particles. Random motions make particles and energy transfer through space. In large systems, the overall effect of microscopic particle motions can translate into a net force at the macroscopic level, if there is not an overall balancing of forces. Because of its probabilistic nature, this force tends to spontaneously offset the very same imbalance that generated it by levelling down and thereby making the distributions of particles and energy throughout the system uniform in relation to the existing force fields. In doing so, the net force can produce work. Once particles and energy have been levelled, no net force is left and the system is then stable with respect to its state functions [8]. Thus, no further work can be done. This is defined as the equilibrium condition, at which macroscopic state functions are invariant. However, as particles motions keep every system in a dynamic equilibrium, macroscopic processes can occur either at equilibrium or non-equilibrium conditions; in the former case they keep the macroscopic state functions invariant and are reversible (and their occurrence is only a theoretical possibility). In the latter case macroscopic processes cause the state functions of the system plus its surroundings to vary and are therefore irreversible (and their occurrence is a spontaneous event). The primary reason for this is, evidently, that irreversible processes are driven by the net force generated by the overall effect of microscopic particle motions - for reversal, this force should be counteracted by another force, but, by definition, once equilibrium is reached no net force is available [8]. As remarkably already noticed by Carnot [11]: "The production of motive power is then due in stem-engines ... to its transportation from a warm body to a cold body, that is, to its re-establishment of equilibrium". In thermodynamics, this phenomenon is described in terms of entropy, the state function that measures the equilibration of a system. When an isolated system is in a state of maximum entropy consistent with its energy (that is, it is at equilibrium), it cannot undergo any further transformation because any transformation would result in a decrease of entropy. Thus, the state of maximum entropy, the equilibrium, is the most stable state for an isolated system [6, 8].

CHAPTER 5

More a Matter of Lost Work Than of Waste Heat

Abstract: It is further highlighted that the concept of work represents a more general way to see what entropy is. That is, the idea of work offers a better framework to understand entropy. A fundamental observation in classical mechanics is that a system is at equilibrium when no work can be done by it. The capability of a system to do work is inversely linked to its stability, which then can be considered to be equivalent to the inertness of the system. Thus, equilibration, stability and inertness are all aspects of the same feature of a system, and such feature is measured in terms of entropy. An equilibrated, stable, and inert system has the highest value of entropy it can reach, and any departure from these conditions results in a decrease of entropy. Therefore, work availability and entropy are inversely linked: the maximal entropy is attained when no work can be done.

Keywords: Classical mechanics, Constraints, Directional macroscopic displacement, Discontinuity, Disequilibrium, Energy availability, Energy dissipation, Free energy, Irreversible process, Levelling of gradients, Lost work, Macroscopic gradient, Overall entropy, Potential energy, Random microscopic motions, Reversible path, Spreading and sharing, Transfer of work, Wasted heat.

SHIFTING THE PERSPECTIVE FROM HEAT TO WORK

The concept of entropy was introduced by Clausius ([3] in English, see the accompanying notes therein for the original German publication; specifically, the word "entropy" was first introduced in 1865) to account for the dissipation of potentially available energy observed in real processes and notwithstanding the first law of thermodynamics implies that energy can neither be created nor disappear into nothing. Indeed, dissipation does not mean that the energy is lost, but, rather, that its availability is lost. In fact, dissipation of energy means that some energy has become no longer available to do work. This is because the system and/or its surroundings are subjected to some transformation that dissipates available energy. The equivalence between a transfer of heat and the production of work in a reversible cycle was indeed the basis of the idea of entropy [3].

In fact, the Carnot cycle shows that a net transfer of heat from hot to cold (where the cold reservoir typically is the environment, *i.e.*, the rest of the universe), which would seem to be a wasting of heat, can be associated with a reversible process. Therefore, a net transfer of heat from hot to cold is not necessarily linked to an overall entropy increase. An overall entropy increase always occurs if the transfer is direct, but if it is mediated by an opportune combination of other physical changes even a process that is ultimately a net transfer of heat from hot to cold can be reversible. It is the using the available energy in full to produce work (which must be used to accumulate some other form of available energy in a way that must be potentially usable to reverse the cycle) that makes the cycle reversible. Any dissipation of the available energy (*i.e.*, any direct, spontaneous transfer of heat) would increase the overall entropy and make the cycle irreversible.

In a reversible process, any change of entropy in the system is balanced by a corresponding, opposite change in the surroundings, typically mediated by a reversible transfer of heat or work. However, most real processes are irreversible and a macroscopic feature associated with an overall entropy increase is that some energy is transformed into a form that is less capable of producing work; it is said that energy is degraded or that available energy is lost. In other words, the energy is conserved (first law), but it undergoes qualitative changes that always decrease (second law) its capability to perform work. Thus, in a process that involves heat and work transfers between a closed system and its surroundings, the overall entropy change is:

$$dS_{tot} = dS_{sys} + dS_{sur}$$

and the entropy change in the system specifically is [12]:

$$dS_{sys} = \delta Q_{actual}/T_{eq} + (\delta W_{rev} - \delta W_{actual})/T_{eq}$$

where δQ_{actual} is the heat actually obtained by the system, δW_{rev} is the work theoretically obtainable by the system by a reversible process, δW_{actual} is the work actually obtained by that system, and T_{eq} is the temperature at equilibration with the environment. Notably, the difference $\delta W_{rev} - \delta W_{actual}$ represents lost work (δW_{lost}). If the process is reversible, the equation reduces to $dS_{tot} = 0$ because the change of entropy in the system is exactly opposite to that in the surroundings ($dS_{sys} = -dS_{sur}$), $\delta W_{lost} = \delta W_{rev} - \delta W_{actual} = 0$, and $dS_{sys} = \delta Q_{actual}/T_{eq}$. However, if the process is not reversible, less work is actually obtained by the system than in a reversible process ($\delta W_{rev} > \delta W_{actual}$), thus that $\delta W_{lost} > 0$ and then $dS_{sys} > \delta Q_{actual}/T_{eq}$. In fact, in an irreversible process, if the heat reservoir guarantees that the temperature of equilibration is the same the system had initially, the lost work is dissipated as heat to the surroundings and the net observed heat transfer to the system is less than the change of entropy of the system:

$$dS_{sys} = \delta Q_{actual}/T_{eq} + \delta W_{lost}/T_{eq}$$

and therefore:

$$\delta Q_{actual} = T_{eq} \cdot dS_{sys} - \delta W_{lost}$$

So, in an irreversible process the overall entropy is increased, that is, some energy is dissipated just in the moving of the process in a spontaneous direction. Therefore, if a transfer of heat occurs from the surroundings to the system, *i.e.* the latter was initially cooler, but the final equilibration temperature, T_{eq}, is considered in the calculations, then Clausius' equation provides an underestimation of the change of entropy in the system. As previously noted, there is a negative equivalence between accumulated work (actually, it is the energy available for doing work that is accumulated) and entropy, and it is now clear that entropy can be seen as lost work (lost in an irreversible process with respect to a reversible one), thus that in an irreversible heat transfer the overall entropy generated because of irreversibility is [12]:

$$dS_{tot} = dS_{sys} + dS_{sur} = \delta Q_{actual}/T_{eq} + \delta W_{lost}/T_{eq} - \delta Q_{actual}/T_{eq}$$

thus,

$$dS_{tot} = \delta W_{lost}/T_{eq}$$

It should be noted that, for the sake of simplicity, the above argument is presented in terms of changes in the internal energy of the system, but it has been shown that a better operational definition is based on the changes in the internal energy of the surroundings [13]. In that definition, $\Delta E = Q + W$ as standard, but Q is the heat effect of the surroundings on the system and W is the work done on the system by the surroundings. In a reversible process the two definitions are equivalent; whereas, in an irreversible one, the work done can be more easily calculated by using the stable and well-defined state functions of the surroundings. In this view, anyway, thermodynamic work should be defined as the negative of the energy change in the surroundings, which is potentially convertible to lifting (or lowering) a mass in the Earth's gravitational field [13].

In general, work is the energy associated with the action of a force at the macroscopic level; specifically, it is the capability of generating force that, when acting on a body, causes displacement at the point of application in the direction of the force. Hence, in its most basic definition work is the capability to displace a body in a specific direction. As seen, a system that is not at equilibrium shows an imbalance of forces, which results in a net directional force and is ultimately headed toward a more balanced and equilibrated state (this will be further discussed); it is such thermodynamic force that pushes the system toward equilibrium and the corresponding work is consequently made potentially available [14]. As this balancing of forces is precisely what characterizes equilibrium, it is clear that work is obtainable from the internal energy of a system only when the latter is not at equilibrium. On the other hand, heat is energy associated with random particle motion at the microscopic level. To pass from random microscopic motions to a directional macroscopic displacement requires some collective action of the particles. This occurs if there is some macroscopic gradient of matter (*e.g.*, ΔP), energy (*e.g.*, ΔT), or other equivalent discontinuity or disequilibrium that can generate and orient the collective action of the particles or a collective transfer of their thermal energy. Work is wasted and energy is dissipated when such gradients are levelled off in a completely irreversible process; if the unbalancing of forces is not exploited to produce work, then it merely provides a spontaneous equilibration toward the direction of the imbalance and the lost work becomes kinetic energy (because of frictions and turbulences) or intermolecular potential energy. This is a general characteristic of spontaneous, irreversible processes and the tendency for levelling matter and energy gradients is just the tendency to spreading and sharing, which is measured by the entropy. The dissipation of a gradient in a process that does not produce any work also illustrates why an increase in entropy is equivalent to wasted work (Figs. **1** and **2**). Indeed, as previously noted for the Carnot cycle, the accumulation of work (as available energy) can counteract the increase of entropy consequent to the thermal gradient dissipation if the work (*i.e.*, the available energy) is stored so that the system is capable of reversing the process. In fact, work accumulation (*i.e.*, accumulation of available energy) means an equivalent reduction of entropy and wasting work equates to an increase in entropy because it corresponds to a levelling of gradients, discontinuities, potentials, and so on, to a more stable condition wherein energy levels are equalized. Energy is conserved, but is no longer available for doing work. As remarked by Strong and Halliwell [15], it is energy redistribution that is significant, not any change in the magnitude of the energy.

A reversible transfer of work can then be seen as a transfer of potential energy, energy that is stored into some gradient or other discontinuity or disequilibrium and that can be exploited to produce work. Clearly, such a discontinuity or disequilibrium must be maintained by some constraint that can be removed to make the potential energy available for doing work. Typically, when a given constraint is removed from the surroundings of a system to ingenerate a disequilibrium between the system and the surroundings it is specifically such a disequilibrium that can, even if very small but prolonged (*i.e.* discrete but quasi-static), induce a significant transfer of work or heat. It can be remarked that even though a discrete, direct transfer of heat from hot to cold is an irreversible process, a discontinuity in temperature can be exploited to produce work; the essential thing is that the transfer of heat cannot be direct but mediated by a properly designed process. This is exactly the way the Carnot cycle works: a net conversion of thermal energy into work is possible exclusively as a result of the fact that a gas at a relatively high temperature can do more work through isothermal expansion than is required for the corresponding isothermal compression at a lower temperature [4].

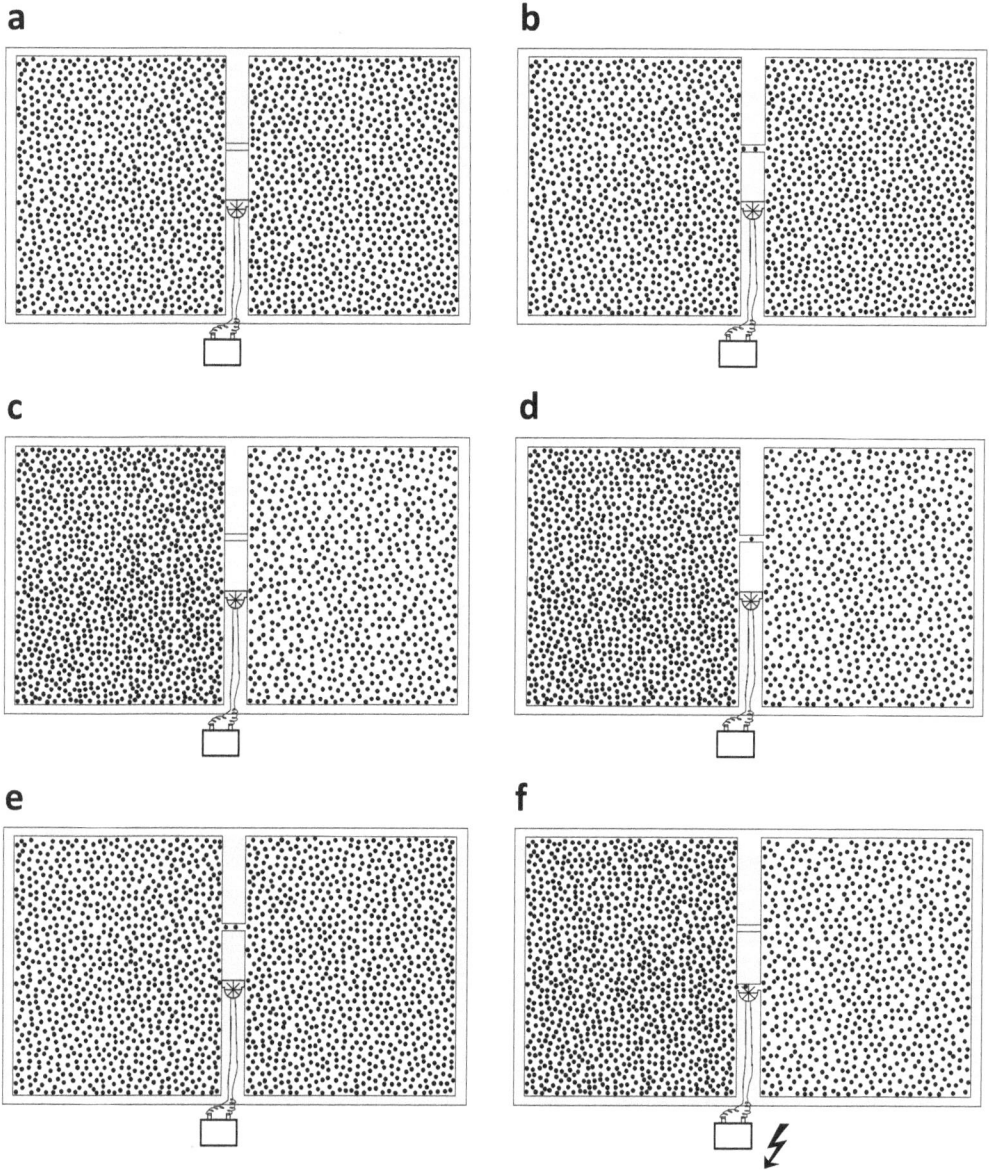

Fig. (2). The dissipation of potentially available energy is a classical mode of increasing entropy. Pressure can be used as an example. Two closed, ideal gaseous systems at temperature $T = 273.15$ K (guaranteed by the surroundings acting as a thermal reservoir) with internal energies E_1 and E_2 are separated by a dividing partition with two tiny holes (whose volumes are considered to be negligible) that can be independently opened and closed. Through each open hole only one particle can pass at a time. In addition, in one hole the passage is free whereas in the other one a hypothetical, microscopic, frictionless turbine is fitted, which reversibly couples the transfer of gas molecules with the generation of an equivalent amount of electricity that can be stored in a perfectly efficient battery outside the systems. In (**A**) the two systems have equal state functions ($V = 22.71$ dm³; 1 mole of gas particles; $P = 1$ bar); both holes are closed. Then, in (**B**) if the two systems are combined by opening the free hole, there is no entropy increase and the combined internal energy is $E_c = E_1 + E_2$. In (**C**), the same systems but with a diverse distribution of the same overall number of particles, where one has its pressure increased and the other decreased by the same amount ($P_1 = 1.2$ bar, $P_2 = 0.8$ bar; then $\Delta P = 0.4$ bar); both holes are closed. When these two systems are combined by opening the free hole, as shown in (**D**), the discontinuity of pressure is immediately evident as an improbable distribution of the gas particles within the combined system, and then it represents the potentially available energy associated with this gradient. In other words, the difference between E_1 and E_2 could be potentially used to produce work when the constraint that separated the two systems is removed. Hence, (**E**) there is a quasi-static, frictionless equilibration that is nevertheless entirely irreversible as no work is produced. The potentially available energy is completely dissipated rather than being used to accumulate an equivalent amount of another type of potential energy like would occur in a reversible process. Since there is a transfer of particles (rather than a mechanical PV equilibration), the potentially available energy arising through the interaction between the combined systems would seem to depend onto the difference between their chemical potentials, but since this diffusive process is associated with a change in the system pressure rather than with an isobaric chemical transformation, the second term of the fundamental equation must be used to calculate the entropy change. Therefore, the overall entropy increase consequent to pressure equalization can be calculated from the volume expansion of the gas corresponding to the exceeding pressure as $\Delta S = nR \cdot \ln(V_f/V_i)$ = 0.4R·ln2 = 2.3 J K⁻¹. If instead, (**F**) only the hole with the turbine is opened in the dividing partition between the two systems having different pressures, so that the microscopic turbine is capable of doing work (this is actually equivalent to the process obtainable by a macroscopic turbine activated by an infinitesimal pressure gradient, as in both cases the pressure differential decreases very slowly), then, at equilibrium, the combined system is the same as in the previous plot (and the equilibrated E_c is also the same), but the potentially available energy is recovered. In fact, the work done consists of a transfer of kinetic energy from the transferring particles to the turbine, and therefore the temperature of the transferred gas would decrease, except that the thermal reservoir keeps it constant. However, the overall entropy has not increased because the original difference of potentially free available energy, $\Delta F = W_{rev} = nRT \cdot \ln(V_f/V_i) = 0.4RT \cdot \ln2 = 629.7$ J, has been preserved, that is, work has been obtained from the levelling of that discontinuity of pressure and it has been used to accumulate an equivalent amount of another form of potential energy in a reversible process (as the energy stored in the battery can be used for operating the turbine to the reverse, to restore the pressure gradient). Preserving the entire originally available potentially energy, even though in another form, makes the process reversible and isentropic. Therefore, the increase of entropy that occurs in the irreversible path (*i.e.* through the free hole) is precisely the dissipation of potentially available energy (whose availability, for this combination of systems, is definitively lost, and this is why the process is irreversible). Thus, in general, unbalanced processes increase the overall entropy because they do not accumulate the whole corresponding potential energy (reversible work) while they dissipate the imbalance (or discontinuity, or gradient, or disequilibrium).

Hence, less heat goes into the cooler reservoir than was removed from the hotter one and the difference is the mechanical work done by the engine. Hence, by combining a series of reversible transformations of the system (in each of which the change does not affect the equilibrium of the forces and then the occurrence of the transformation is only a theoretical possibility), obtained through the theoretical constraints of some state function and the unconstraining of others, the system can be left unaffected (as the series of transformation is cyclical), but eventually a transfer of heat is transformed into work. In this way, assured that the thermal reservoirs remain unchanged, the temperature difference represents a form of persistently available energy. In most practical situations, it persists until there is fuel to release heat and the environment allows efficient cooling.

The subsequent reversible equilibrations of the system to different temperatures by means of adiabatic changes of volume shows the connections among state functions can be exploited to make what would otherwise be an irreversible transfer a reversible process. In general, there can be a reversible indirect path for the loss of a gradient (of temperature, pressure, concentration, or other potential energies) even though a direct transfer would represent an irreversible path for the same process. The key point is that in a reversible path the potential energy is simply transferred from a given kind of energy to another exactly equivalent type. Although a fully reversible process is only theoretical, a quasi-static one can provide a path that has a very low degree of irreversibility (*e.g.*, in practice, heat can be transformed into work, but never completely, which is another way to formulate the second law of thermodynamics).

It can now be noticed that work can be done when there is some energy that is potentially available to do it. Potentially available energy, in other terms, can give rise to a transfer of work when a process is reversible, or no work at all if a process is entirely irreversible. Any path that has an intermediate degree of irreversibility would result in an intermediate proportion of work produced.

When two systems, each having its own given internal energy but also characterized by different levels of some kind of potential energy, are combined, then the combined system has an internal energy that is the sum of the energy of the two original systems, but the difference in the potential energy can be used, or "extracted", to obtain work (Fig. **2**). Indeed, work can be "extracted" only during the loss of a gradient and the reversible process represents the upper limit of the "extractability" of work in a process[10]. Since this part of the internal energy can be "extracted", that is, it is available for doing thermodynamic work, it is also called "free energy" [4] since it is freely available to perform some kind of work. There are actually numerous different ways of calculating available energy because the kind of work that can be done depends on which state functions are fixed [9]. Gibbs and Helmholtz free energies are the most important ones, as they are suited to conditions that are approximated in many common real systems. Notably, in closed systems it is just the internal energy of the system (in chemistry this is usually intended to mean the thermal energy of the system) that represents the form of energy freely available to perform work.

Any "extractable" work that is not "extracted" during these processes cannot be further "extracted" from the internal energy of the combined system once this has reached equilibrium, *i.e.* when the availability of the energy has been dissipated. Clearly, some internal energy of the system can be subsequently "extracted" if the equilibrated system is further combined with another system differing in the same or a different level of potential energy. The levelling of the potential energy associated with a gradient represents an increase in entropy and if no work is obtained then the process is entirely irreversible (Figs. **1** and **2**). The internal energy of each original system is conserved, but the potentially available energy is no longer available for doing work, *i.e.* it has been dissipated. Since entropy measures the intensity of "dissipation" of energy, the entropy is what determines the proportion of internal energy that is "free" to do work. Therefore, when the entropy of a system increases, the proportion of its internal energy that is available as "free energy" decreases.

So, a direct transfer of heat to a system is an increase in the entropy of that system because it reduces the temperature gradient with the surroundings and thus the available potential energy associated with such a gradient. It is dependent upon the path whether this results in a corresponding decrease in entropy for the surroundings (reversible path) or in an overall increase of entropy (irreversible path).

The concept of work is also important in the classical view of entropy because it is at the basis of the idea that a reversible process can only occur through a series of equilibrium states. In fact, if the series of equilibrium states is not continuous[11], then there must be some jump in the change of the state functions involved in the process (Fig. **1**). Any jump implies that the balance between the system and its surrounding has given up, even if minimally. By that, the close connection between the state functions of the system and of its surroundings undergoes a corresponding interruption and therefore the transfers of internal energy (Q + W), which, under equilibrium conditions, mediate the equivalent changes of the state functions occurring in the system and its surroundings, result decoupled from such changes. This indicates that the jump was not linked to work done or received by the system and work was then not accumulated (as potential energy) for reversing the process. In other words, there has been some free expansion, spontaneous chemical reaction, or other change that was not accompanied by a corresponding accumulation of potential energy that could be used to make the process fully reversible[12]. Each jump is therefore a lost opportunity to store energy available for work and is directionally driven[13] by some overall increase of entropy (Fig. **1**)[5].

Even in the instance of a falling ball, the idea that the waste heat produced does not cause a temperature increase and that therefore it represents a discrete increment of the entropy of the universe exactly equivalent to the released heat, though correct, is a little unsettling as conceptual approach. It is much easier to appreciate that the gravitational potential energy that is dissipated once the ball has fallen represents an increase of entropy equivalent to the lost work. Sometimes, the concept of waste heat feels a bit contrived if taken by itself, without reference to the lost work. Since they are equivalent and consequent, the best way to conceptualize waste heat (as well as free expansion and free diffusion) is seeing it as a common way of losing work.

Briefly, classical mechanics, which was mainly concerned with the reciprocal transformation of heat and work, improved our understanding of entropy by showing that the notion of entropy is better expressed as lost work than as waste heat and the former is a more general explanation of entropy.

CHAPTER 6

Entropy in Statistical Mechanics

Abstract: The systems we can directly see are composed of huge numbers of particles. So, the properties of these systems are obtained as statistical averages of the effects of their particles. This casts a conceptual bridge between the macroscopic world wherein we observe systems with their overall properties and the microscopic world where particles with their own properties dominate the scene. Statistical mechanics shows that the former world is determined by what happens in the latter, and this approach provides a better, finer understanding of what's going on at the macroscopic level and why.

Keywords: Accessible microstates, Boltzmann entropy, Constraints, Equilibration, Gibbs entropy, Inertness, Macroscopically stable equilibrium, Maximum entropy, Maxwell-Boltzmann distribution, Microstates, Microscopically dynamic equilibrium, Phase space, Second law, Spreading function, Stability, Statistical mechanics, Thermodynamic ensemble, Trajectory.

THE CONCEPT OF "MICROSTATES" AIMS TO DESCRIBE ENTROPY IN TERMS OF THE ACTUAL CHANGE THAT OCCURS IN THE SYSTEM WHEN SOME PROCESS MODIFIES IT

Although Clausius' equation is the working definition of entropy, even in classical mechanics, and the equivalence of entropy with lost work is very useful for measuring entropy changes, both these approaches do not provide a clear understanding of what entropy is by itself. Subsequent theoretical studies were therefore undertaken by statistical mechanics (which deals with the microscopic details of macroscopic systems) that showed entropy is an extensive property of a macroscopic system (*i.e.* a system that includes a large number of particles) that

Alberto Gianinetti

derives from the statistical features of the ensemble of microscopic states (microstates) the system can assume. Microstates are different states, detailed at the microscopic (particle) level, that a system can attain at given values of its macroscopic state functions (like N, E, and V, number of particles, energy, and volume, respectively; that is, the state functions that characterize isolated systems, which are the most fundamental thermodynamic systems to study). Each microstate is defined by both the way N particles are physically arranged in the system volume and how their different levels of thermal energy (which are exactly proportional to their velocities if all the particles are identical) are distributed among the particles. Formally, the microstates are said to be points in a phase space, whose coordinates define the spatial localization and the energetic level of each single particle of the system [9, 17]. In other terms, we define the phase space of a system as the multidimensional space of all of its microscopic degrees of freedom [9]. Thus, for classical systems, the positions and momenta (*i.e.*, instantaneous movements) of all constituent particles describe each microscopic state. If the particles of the system are all ideal and spherical, the multidimensional phase space has a number of dimensions that is given by the number of particles multiplied by six (namely, three positional coordinates plus three momentum coordinates, one for each spatial direction). This can be thought of as specifying a point in a $6N$-dimensional phase space, where each of the axes corresponds to one of the momentum or position coordinates for one of the atomic particles [9].

An ensemble of microstates is a collection of the various microscopic states of a system that correspond to the single macroscopic state of that system, which is characterized by a given set of state functions. It should be noted that an ensemble is the theoretical set of microstates that the system can assume over time given its overall thermodynamic properties, which means that, in actuality, these microstates are not independent through time (as if they were coloured balls extracted from an urn). They necessarily constitute a time sequence because each microstate is causally determined by what the particles were doing in the microstate at the previous instant. In other terms, over time the system moves through accessible microstates along a trajectory in the $6N$ phase space that reflects the ceaseless changes at the microscopic level as particles continuously move, collide, and exchange energy. The trajectory is deterministic, that is, we are guaranteed that all systems that sit at the same coordinates in phase space will evolve in exactly the same way [9]. Nonetheless, in the long run, it is expected that all accessible microstates are visited with frequencies that are (approximately) equal because they have the same energy (in an isolated system) or approximately the same energy (in a system equilibrated at a fixed temperature). In fact, although we cannot exclude the fact that the deterministic trajectory is trapped in a vicious circle that leaves out some, or many, of the theoretically equiprobable microstates, the huge number of particles included in a macroscopic system and their consequently immense number of interactions makes this possibility very remote. Thus, an implicit assumption of statistical thermodynamics is that for any isolated system, devoid of physical constraints, there must be full actual access to all the theoretically accessible microstates having the same, given energy, E, and conforming to other thermodynamic constraints, like V, N and the probabilistic distributions of matter and energy across every macroscopic parcel of the system. In other words, eventually the trajectory in phase space must visit all the points in the region representing the accessible microstates of the system with equal frequency [17]. Actually, this is an assumption of perfect randomness of the accessible microstates in the very long run, which is expected to hold whether the universe is either deterministic (that is, the state of the universe along time is fully determined by its initial state) or not; the mixing of the particles is expected to be complete (*i.e.*, with no theoretically accessible combinatory arrangement left out) in any case.

On the other hand, since a microstate's occurrence is not independent through time, some microstates are most probably not accessed because they would require trajectories that lead toward levels of higher and higher available energy (that is, against the increasing intensity of a net force that tends to restore the equilibrium) and then these trajectories have a higher and higher probability of reverting to a uniform energy level. For example, consider the case where most of the particles of a gaseous system would happen to move toward an edge of the system space: this is an instantaneous event that can casually occur; even though it has a very low probability, it is equally probable with respect to other random microstates. If, however, this synchronous behaviour should last for many instants, the corresponding sequence of microstates would involve a progressive spontaneous concentration of particles in that part of the system. This situation would no longer be equally probable with respect to accessible microstates, rather, at equilibrium, an inhomogeneous microstate has a probability tending towards zero, that is, it is not really accessible, or, better, it is not actually accessed. Formally, we can distinguish between microstates that are not accessible given the state functions of the system (*i.e.*, by hypothesis) and microstates that are actually accessible according to the first law, but access is extremely improbable according to the second law. In other words, since matter, energy, and space neither appear from nor disappear into nothing, microstates that do not correspond to the given values of these state functions are not possible, whereas microstate that would be possible according to these constraints can further be distinguished by their probability of occurrence as either probable, if they do correspond (at least quite closely) to equilibrium conditions, or very improbable, if they significantly deviate from equilibrium. How do we establish the equilibrium conditions and measure eventual deviations from it? By means of entropy, which is primarily a measurement of the net force present in the system that tends to restore the equilibrium.

THE BOLTZMANN ENTROPY

Thus, for an isolated system (*i.e.* one that does not exchange energy or matter with its surroundings) at equilibrium the entropy can be calculated, for an ideal gas, by the Boltzmann expression, which is defined [9] in terms of the probabilistic arrangement of the system particles, as being proportional to the logarithm of Ω, the number of microstates that a gas can occupy:

$$S(E, V, N) = k_\mathrm{B} \ln \Omega$$

where k_B is the Boltzmann constant. In other words, classical entropy is a statistical function of all the microstates in which the system particles can be arranged. More specifically, for a given set of macroscopic state functions, the entropy measures the probability that the system has spread out over the different possible microstates (that is, over the whole magnitude of the ensemble), where a microstate specifies all the molecular details about the system, including the position and velocity of every molecule. The basic idea behind this definition is that the larger the number of microstates over which the system can fluctuate, the more equilibrated, stable, and inert the system is overall, so that entropy is a measurement of thermodynamic equilibration, stability (at the macroscopic level), and inertness. In fact, an isolated system is equilibrated when there is no net internal force that can modify it and therefore is stable, *i.e.* its macroscopic properties do not change with time. At equilibrium, an isolated system is also inert, just because there is no force that can do work. A force would occur only if there were energy available to be extracted, that is, a gradient/discontinuity/ inhomogeneity in the system that would make it depart from the equilibrium. Equilibration, stability, and inertness are hence all different aspects of the same physical condition. This is why I suggest that a useful definition of entropy is "a function of the system equilibration, stability, and inertness". We will see that even the maximization of the number of accessible microstates is still another facet of the very same condition.

A logarithmic function is necessary to link the fact that entropy is an extensive property (that is, if there are two systems with the same temperature, pressure and composition, their entropies sum when the systems are connected to form a combined system: $S_{1+2} = S_1 + S_2$) to the fact that the number of microstates (Ω) in the combined system is the product of those of the two separate systems ($\Omega_{1+2} = \Omega_1 \cdot \Omega_2$; as each of the Ω_1 microstates of the one system can combine with each of the Ω_2 microstates of the other system); in fact

$$S_{1+2} = S_1 + S_2 = k_B \ln \Omega_1 + k_B \ln \Omega_2 = k_B \ln (\Omega_1 \cdot \Omega_2)$$

As seen, one of the fundamental assumptions of statistical mechanics, idealized by the formulation of the Boltzmann entropy, is that for an isolated system at equilibrium all the microstates having energy equal to the energy E by which the system is characterized are equally likely to be occupied. This is called "the assumption of equal a priori probability". For systems at fixed values of macroscopic state functions N, E, and V, the probability of the microstates is therefore the same throughout the whole ensemble (such isoenergetic ensemble is traditionally defined "microcanonical").

THE GIBBS ENTROPY

Realizing that the diverse microstates can actually have different probabilities of occurrence (as the equilibrium of a system is given by a perpetually dynamic equilibrium of its parts rather than a fixed one; thus fluctuations are continuously occurring), particularly in systems that are neither isolated nor at equilibrium, Gibbs [18] formulated a more general equation:

$$S = -k_B \sum_m \wp_m \ln \wp_m$$

where $\wp_m$ is the probability for the system to be in the microstate m over an ensemble of Ω accessible microstates, and, of course:

$$\sum_m \wp_m = 1$$

that is, these probabilities must sum up to one. This equation is applicable to isolated systems, as well as to close and open systems, even far from equilibrium conditions. Actually, the Gibbs entropy formula, which defines the sum is over all microstates of the system, gives the entropy for any statistical mechanical ensemble [9]. As we will see, for an isolated system (with N molecules, volume V, and internal energy E) in equilibrium conditions, the Gibbs equation simplifies (with a good approximation) to $S(E,V,N) = k_B \ln \Omega$, which is the Boltzmann equation for an ideal gas at equilibrium. In fact, at equilibrium, if all the microstates have equal (uniform) probabilities, the probability of each one of the Ω microstates is $1/\Omega$, as their overall sum must be one. Therefore, Gibbs entropy reduces thusly:

$$S = -k_B \sum_m 1/\Omega \ln 1/\Omega = -k_B \, 1/\Omega \sum_m (\ln 1 - \ln \Omega)$$
$$= -k_B (1/\Omega \sum_m \ln 1 - 1/\Omega \sum_m \ln \Omega) = -k_B (0 - 1/\Omega \sum_m \ln \Omega)$$
$$= -k_B (- 1/\Omega \sum_m \ln \Omega) = -k_B [- 1/\Omega (\Omega \ln \Omega)]$$
$$= k_B \ln \Omega$$

This is the maximum entropy attainable by the system, and indeed the second law of thermodynamics, in one of its alternative formulations, states that an isolated system at equilibrium will maximize its entropy [9].

The Gibbs equation highlights that entropy is a probabilistic phenomenon, but at first sight this appears difficult to combine with the original idea of its equivalence with heat transfer. This point can be made clearer by viewing entropy as a probabilistic tendency to spread; entropy can be considered a "spreading function" [19, 20], *i.e.* a function that assures an equitable distribution of energy, and eventually particles, between adjacent, non-isolated systems (which, in terms of the Gibbs formula, tend to have equally probable distributions over microstates). Accordingly, the principle of entropy's increase (the second law of thermodynamics) is interpreted as nature's tendency towards maximal spreading and sharing of energy [19].

This includes, for example, both the equalization of pressure between open systems (as occurs for a gas, by particle spreading from higher to lower pressure), as well as the adjustment of the distribution of energy through particles at different temperatures. The Maxwell-Boltzmann distribution, for example, shows that particles must be distributed in velocity at a given temperature, with a very specific form for that distribution [9]. In fact, isolated systems equilibrated at different temperatures show diverse distributions of quantum states,[14] approximated by the Maxwell-Boltzmann distribution of thermal energies:

$$p_j = n_j/N = e^{-\beta \epsilon j}/\Sigma_j\, e^{-\beta \epsilon j}$$

where $p_j = n_j/N$ is the expected probability to find n_j of the total N particles of the system having thermal state j of energy $\epsilon j = \epsilon 0 \leq \epsilon 1 \leq \epsilon 2 \ldots$, for a given value of β, that is, for a given temperature T, as [21]:

$$\beta = 1/k_{\mathrm{B}}T$$

Higher temperatures equate to greater spreading among thermal states. According to the Gibbs formula, this contributes to increasing entropy[15].

CONSTRAINTS LIMIT SPREADING AND THUS REDUCE ENTROPY

Spreading to more equitable conditions increases entropy. However, spreading can be hindered or prevented by some constraints. A constraint occurs when the system does not have full access to all possible internal configurations (microstates), whether the constraint is physically imposed or naturally results from slow system kinetics (the latter would be a restraint, which can be considered a constraint if the kinetics are so slow as to be negligible). In this sense, a constraint can be seen as a fully effective and stable restraint, at least relative to the duration of a given process. Typically, an internal constraint involves some perturbation away from equilibrium, and such a perturbation persists until the constraint persists. For example, a transient increase in the density or energy in some spatial region can be a perturbation, but it does not persist in the absence of specific constraints. At a microscopic level, constraints reduce the number of available quantum levels, spatial and/or thermal, for the particles. In other words, they reduce the number of accessible microstates; therefore, the fewer the constraints on how and where the particles can move, the greater the entropy [22]. Accordingly, releasing any degrees of freedom associated with a constraint will always maximize and increase the entropy [8, 9]. For this reason, and with the perspective that "more entropy means more microstates", adopting "freedom" as a metaphor for entropy has been proposed [23].

Gravity is one kind of constraint: clumping of matter by gravitational forces caused the formation of galaxies and stars and this means that clumps of hot matter were spontaneously formed in a landscape of cold space, almost devoid of matter and energy. This seems opposed to a uniform distribution of matter and energy consequent to spreading and sharing, but this function is always conditional with the existing constraints. The presence of gravitational forces is just a powerful constraint, capable of substantially modifying the equilibration of matter distribution. Analogously, the existence of atoms, which are concentrated clumps of energy (as $E = mc^2$), is caused by electromagnetic, and strong and weak nuclear interactions, which oppose energy spreading. In most everyday systems, however, only thermal energy is free to spread and it is the part of the internal energy of a system that is directly proportional to the temperature. The rest energy, $E = mc^2$, is therefore not considered in common systems and it becomes relevant, and must then be considered, only when nuclear reactions are studied.

In general, fields of forces can alter the probabilities of particle arrangements, for example, in the case of gravity and the Earth's atmosphere (the atmosphere is a system of gases that do not diffuse toward the interplanetary space because the Earth gravity keeps them stuck to the Earth, thereby generating a gradient of density that decreases with altitude) or of an electromagnetic field, as if the particles have a net charge, like ions, their arrangement can be strongly affected by such a field[16]. In these cases, the spatial distribution is not uniform, although it is the most probable arrangement.

Succinctly, mechanical statistics shows that a macroscopically stable equilibrium (characterized by invariant state functions) corresponds to a microscopically dynamic equilibrium with the system that fluctuates over a huge number of microscopically distinct states compatible with that macroscopic equilibrium. It also shows that, in accordance with statistical considerations, the most stable state of an isolated system (*i.e.*, the equilibrium state) must be the state of highest probability consistent with the given total energy of the system [6].

CHAPTER 7

The Effect of Temperature

Abstract: Since the features of macroscopic processes derive from the microscopic behaviour of the system's constituent particles, it is shown that temperature, which appears to us as a simple property of macroscopic bodies, is indeed a complex effect of the movement of the body's particles.

Keywords: Boltzmann factor, Dominating distribution, Heat transfer, Increase of entropy, Maxwell-Boltzmann distribution, Non-uniform energetic distribution, Partition functions, Peaked probability distribution of thermal energy, Probabilistic equilibration, Redistribution of thermal energy, Reduction in entropy, Spontaneous change, Uniform spatial distribution.

The entropy of a system increases progressively more slowly (logarithmically) as the number of accessible microstates grows with temperature[17]. So, entropy shows smaller incremental increases as temperature grows. It is null at absolute zero, at least for perfect crystals, which is a common way to state the third law of thermodynamics.

A discrete, irreversible transfer of heat causes a change of temperature and then influences the distributions of thermal energies among the particles of a finite system and of its surroundings (unless they can be considered as infinite) and always results in entropy increase. Since entropy shows reduced increases as temperature grows, an increase in temperature of the coldest between the system and its surroundings determines that the increase of its entropy is greater than the reduction in entropy due to the decrease in temperature of the warmer one. As seen, in an irreversible heat transfer, Q directly flows from a hot body at T_H to a cold one at T_C and the respective gain and loss of entropy for the two bodies are -

$\delta Q/T_{\mathrm{H}}$ and $\delta Q/T_{\mathrm{C}}$, respectively, and therefore, as $T_{\mathrm{H}} > T_{\mathrm{C}}$, the overall entropy change is:

$$\mathrm{d}S_{\mathrm{tot}} = \delta Q/T_{\mathrm{C}} - \delta Q/T_{\mathrm{H}} > 0$$

Thus, the transfer of heat from hot matter to cold is a redistribution of thermal energy among particles that can be seen as being driven by an overall increase of entropy. In classical thermodynamics this was the only way to account for the universality of the phenomenon by which heat transfers from hot to cold bodies. Indeed, it is precisely to justify why such a kind of process is spontaneous and irreversible[18] that the concept of entropy as a state function of a system was introduced. Thus, to assert that heat spontaneously transfers from higher- to lower-temperature bodies but never in the reverse direction is just another way to state the second law of thermodynamics. It turned out that an increase of entropy is the drive for any heat transfer as well as a natural tendency that determines many other thermodynamic features of the universe. Indeed, a common formulation of the second law of thermodynamics is "the entropy of an isolated system increases in the course of a spontaneous change: $\Delta S_{\mathrm{tot}} > 0$" [25]. In other words, any spontaneous change is always associated with an increase of entropy, and the entropy change is therefore the natural function used to assess the spontaneity of every physical change[19].

Nonetheless, a non-intuitive point in interpreting entropy as a "spreading function" is precisely the increase of entropy with temperature. On the one hand, spatial spreading is an easily comprehensible concept, since diffusion commonly acts to smooth gradients, and this fits well with the idea that maximum entropy of an isolated system occurs when all the microstates have the same probability (that is, they are probabilistically equilibrated to a uniform spatial distribution [8]). On the other hand, however, it is far less clear why the spreading over quantum states of higher energy that occurs when temperature is increased should be a natural tendency. In this case, the idea of "spreading function" appears especially puzzling because, at the equilibrium (at a given temperature), the system does not have a uniform distribution through quantum states (that is, they are probabilistically equilibrated to a non-uniform energetic distribution). In fact, in the case of thermal energy, the equilibration is not to a uniform distribution of the frequencies of particles among energetic levels, but to the Maxwell-Boltzmann distribution, which is peaked. Thus, explaining entropy in terms of spreading and sharing [19, 20] turns out to be a very efficacious metaphor for spatial interpretations, whereas it is more difficult to convey when more nuanced and circumstantial interpretations of "spreading" and "sharing" are intended [26].

In contrast with space, over which a uniform distribution of the particles seems evidently the most probable condition (although gravity affects the spatial distribution of matter), the average thermal energy is constrained to $k_B T$ and therefore its distribution among molecules has to have an average of $k_B T$. This is a key constraint and is related to the first law, as it implies that the internal energy of a system is conserved and the average kinetic energy of the particles is therefore fixed. In addition, the distribution of energy is established by how it can be partitioned over the particles, *i.e.* using the thermal, or momentum, partition function. In statistical mechanics, partition functions are equations that assign probabilities to microstates, in relation to space and energy, and normalize such probabilities so that they sum to one over all microstates [9]. The momentum partition function is the denominator of the Maxwell-Boltzmann distribution:

$$\Sigma_j \, e^{-\beta \varepsilon j}$$

where $\varepsilon j = \varepsilon 0 \leq \varepsilon 1 \leq \varepsilon 2 \ldots$, is the energy of thermal state $j = 1, 2, 3, \ldots$, and $\beta = 1/(k_B T)$. This expression defines the way the thermal energy of the system, $k_B T$, is probabilistically partitioned over the different possible thermal states and energy levels[20] in terms of the fraction of particles that occur at each given thermal state[21]. As the momentum partition is a negative exponential function of the energy of the thermal states, it assigns decreasing probabilities to levels of higher energy. This is a second, fundamental restraint on the energy distribution. On the other hand, higher temperatures mean a more even distribution across thermal states, particularly at higher ones. Thus, there is a competition between the number of thermal states, which tends to grow as the state's energy grows, and the probability of higher-energy quantum states, given by Boltzmann factor $\exp(-\varepsilon j/(k_B T))$, which decreases at higher energies of these quantum states (*i.e.*, these states tend to be less populated because of their high energy). In other words, the fundamental distribution of particles to thermal levels is not equal, but probabilistically equilibrated.

Still, however, it is not intuitive why in the spreading over quantum states of higher energy that is consequent to a discrete transfer of heat to the system, which thereby increases its temperature, the gained thermal energy turns out to have an effect on the capability of the system to perform work (and thereby on the decrease of its inertness/entropy) that is subdued with respect to what it would have been if the heat transfer were isothermal. The answer lies in the properties of the Maxwell-Boltzmann distribution.

As seen, the Maxwell-Boltzmann distribution of thermal energies follows the equation:

$$p_j = n_j/N = e^{-\beta \varepsilon j}/\Sigma_j \, e^{-\beta \varepsilon j}$$

which shows that there is an exponentially decreasing probability for levels of higher energy, but also a constraint on the overall thermal energy of the system to $k_B T$, and thus this probability distribution is peaked. The momentum partition function shows how, at a given temperature, the thermal energy of the system is partitioned over the different thermal states. Thus, the Maxwell-Boltzmann distribution expresses the most probable distribution of molecules over the available energy levels[22] [25]. As this distribution is not uniform (that is, the frequency of particles with a given level of thermal energy is diverse for different energy levels), the equilibrium distribution of thermal energies among the particles of the system corresponds to different numbers of particles for the diverse levels of thermal energy. That is, the probability of having each particle at its actual thermal energy is balanced for such energy over the entire number of particles and thus altogether they are in an equilibrated condition. Thence, when the temperature is raised, the energy is rearranged among the particles to fit with the most recent equilibrated distribution and in doing so it dissipates part of the thermal energy (as the system settles into a more probable state). It is as if some available energy were lost in this equilibration process, even though it is at present difficult to say how useful work could have been directly extracted from the disequilibrium in the distribution of thermal energies. So, while the system receives some heat and thus its temperature is increased, the distribution of thermal energy resettles to a new equilibrated condition, therefore reducing the efficacy of the heat gain in terms of utilizable work (this shows that the equilibration in the distribution of thermal energies is indeed a loss of work). This phenomenon explains why there is a less than proportional gain in the capability of the system to perform work following a discrete irreversible heat transfer: some energy is dissipated in the adjustment of equilibration through energy levels and thus the overall entropy has increased[23]. In general, when a discrete amount of heat is transferred to a body that thereupon changes its temperature, there is an adjustment in the Maxwell-Boltzmann distribution of thermal energies among particles that also causes an increase of entropy, which results in a greater inertness. Thus, heat spontaneously transfers from warmer to cooler matter just because thereby it increases the overall entropy (of the system plus its surroundings), that is, it increases the equilibration, stability (at the macroscopic level), and inertness of the system plus its surroundings. In addition, in most real systems, some thermal energy is spent to weaken interparticle bonds, and thus it is further dissipated and entropy is further increased.

As already noted, although the thermal, or momentum, partition function shows exponentially decreasing probabilities for levels of higher energy, the distribution of energy is sharply peaked around the mean energy. This is because the continuous collisions of the particles cause the thermal energy to promptly and continuously redistribute and equilibrate among particles. There will therefore be only a few particles with either very high or very low energy with respect to the mean. Accordingly, the particles roughly tend to have similar energies and the overall energy must sum up to the level given by $k_B T \cdot N$. Therefore, the mean energy of most particles is necessarily the most probable one, *i.e.* the one around $k_B T$. Thus, the thermal partition function gives a peaked probability distribution of thermal energy among the particles because there is an exponentially decreasing probability for levels of higher energy plus a constraint on the overall thermal energy of the system to $k_B T$. As all the instantaneous actual energy distributions (representing the energy component of the canonical ensemble, wherein T, V, and N are fixed[24]) have approximately the same energy (since they are constrained to $k_B T$), the restraint to access very high energies is strong. In addition, as the continuous collisions of the particles make the redistribution of energy a process highly efficient and then practically unfailing, the Maxwell-Boltzmann distribution is the dominating distribution of the thermal energy of the system observable through time, and therefore over members of the ensemble.

CHAPTER 8

From Nature's Tendency To Statistical Mechanics

Abstract: Some processes happen spontaneously. What, at a macroscopic level, appears as a nature's tendency, is an effect of the complex statistical behaviour of the microscopic particles: their overall net effect emerges at the macroscopic level as a spontaneous force that determines if and how a system can spontaneously change, and if and toward which direction a process is therefore started.

Keywords: Canonical ensemble, Canonical partition function, Crossproduct, Momentum partition function, Phase space, Positional partition function, System configuration, Tendency.

A couple of questions that arise from referring to the entropy increase as a "tendency" are: (i) a tendency toward what?, and (ii) what causes the tendency? As already established, the tendency is toward a more equitable states (however, since the Gibbs expression specifically deals with probabilities, I shall use "equiprobable", *i.e.* equally probable, instead of equitable), but what thrusts matter and energy to reach the condition of equiprobable states? To get the answer to this question requires further delving into the meaning of entropy, and particularly, of microstates. Each microstate is defined by both the way the N particles are physically arranged in the system volume and how their different levels of thermal energy (which are exactly proportional to their velocities, if all the particles are identical) are distributed among the particles. In other words, they are points in a multidimensional phase space, whose coordinates define the spatial localization and the energetic level of each single particle of the system [9, 17]. The phase space is therefore the crossproduct (*i.e.* the mathematical intersection) of a positional partition function (a function that describes the physical arrangement of the particles in the system volume) by a momentum partition function (a function that describes the distribution of the particles over the accessible momenta, in practice, their thermal levels). In the overall partition of energy and space the number of microstates is therefore:

$$\Omega = \Omega_m \cdot \Omega_p$$

where Ω_m is the number of accessible momenta and Ω_p is the number of accessible positional, *i.e.* physical, arrangements of the particles [9]. Together, the thermal and positional distributions define what is called the (microscopic) configuration of the system. A configuration is the set of microstates that have (approximately) identical distributions of energy. We will see that this definition is important because the assumption of equal *a priori* probability (approximately) holds even for a canonical ensemble, that is, when just the temperature, rather than the energy, of the system is fixed.

It is worth noting that the mathematical intersection of the positional and energy distributions makes always reference to the most probable distributions for both aspects, that is, the uniform spatial distribution of particles (eventually modified according to a gravitational field or other constraint) and the Maxwell-Boltzmann distribution of particles over levels of thermal energy. Microstates that deviate from these distributions represent, by definition, less probable microstates (as expressed by the Boltzmann factor in the respective partitions). Hence, the probabilities of the Ω_m accessible momenta are not uniform, but are given by the Maxwell-Boltzmann distribution. Correspondingly, the number of accessible momenta, Ω_m, and therefore the overall number of accessible microstates, $\Omega = \Omega_m \cdot \Omega_p$, are always constrained, and therefore limited, by the Maxwell-Boltzmann distribution. In fact, the momentum of each particle is not independent of that of the other particles as altogether they are constrained to E in an isolated system (or to E^*, if T is fixed, as we'll see). Simultaneously, the arrangements of the particles must ensure a uniform density distribution across the system volume. In other words, the crossproduct $\Omega = \Omega_m \cdot \Omega_p$ does not include the complete set of all the mathematically possible combinations. It should also be reminded that, at each instant, the distribution of particles among thermal levels can eventually move away from the Maxwell-Boltzmann distribution, just as the spatial arrangement of the particles can deviate from a uniform distribution, but the overall amount of thermal energy is in any case constrained to $k_B T$ (given the system temperature), just like the number of particles is constrained to N, at least in an isolated system.

Specifically, three possible kinds of deviations from the most probable configuration can occur, for every kind of system: (i) deviations from a uniform distribution of the particles throughout the system (like differences in the pressure of a gas over the internal space of its container), *i.e.* non-uniform positional arrangements; (ii) deviations from the Maxwell-Boltzmann distribution (like an increase of the frequencies of particles at low and high energy levels, which must anyway be symmetrically weighted to leave $k_B T$ unchanged); (iii) an interaction effect, that is, a non-uniform distribution of energy levels throughout the system volume (like inhomogeneities in the temperature of a gas over the internal space of its container). These deviations can either be casual fluctuations or they can be due to constraints. In any case they affect the entropy of the system and therefore, for a system at a given temperature, are considered in the canonical partition function.

The canonical partition function assumes that the microstates over which the macrostate of a system can fluctuate are constrained to T, V, and N, that is, the temperature, the volume, and the number of particles are exactly defined. It is typically assumed that the system is closed, *i.e.* there is no exchange of particles, but it is in contact with a very large bath (the surroundings) that keeps the temperature constant and with which the system can exchange thermal energy. Hence, there can be some fluctuations in the energy of the system that derive from ephemeral transfers from/to the surroundings (so, the macrostate of the system is considered to have temperature T, but it can instantaneously fluctuate over microstates that can deviate from the thermal energy strictly corresponding to T). Actually, the idea is that the surroundings (the bath) is fixed at a given temperature T (and small transfers of energy from/to the system do not affect its temperature because the bath is very large with respect to the system), whereas the system has its energy at equilibrium with the surroundings, but its temperature is not perfectly fixed. The temperature of the system is just kept at T by the equilibration consequent to the exchange of thermal energy with the bath. Indeed, these theoretical conditions are aimed at studying thermal equilibrium. In this sense, it can be useful to remind the reader that whereas the energy of the system can be directly computed as the sum of the energy of all its constituent particles, because the internal energy is additive, the temperature is only a macroscopic function statistically derived from the average velocities of all the particles, so that only an infinitive bath can be said to have exactly a given temperature. This is why the temperature of the bath together with the equilibration of the system with the bath is considered in defining T, rather than just the temperature of the system.

In this way, the canonical partition function considers that the diverse microstates can have different probabilities (as noted when the Gibbs entropy was introduced), and therefore they are weighted for their respective probability according to the Boltzmann factor, $e^{-\beta E_m}$, where E_m is the energy of microstate m. The Boltzmann factor then gives the relative probability of each microstate, *i.e.* the frequency with which the system visits it, when a system is at a constant temperature rather than constant energy [9]. This is exactly the same concept that the Maxwell-Boltzmann equation is based on, but it is applied here to the energy of each microstate in the ensemble representing the system, rather than to the energy of each particle of the system. Hence, the system will visit with substantial and equal frequency all the accessible microstates with energy $E_m = k_B T$, but the visiting frequency drops off quite rapidly for $E_m \neq k_B T$ [14]. Indeed, the Boltzmann factor is central to the conceptualization of entropy. In statistical mechanics, the canonical partition function is then defined as[25]:

$$Q(T, V, N) \equiv \sum_m e^{-\beta E_m}$$

that is, it is calculated as the sum of accessible microstates weighted for their respective probabilities; it works as a normalizing factor for the microstate probabilities, *i.e.*, for calculating the probability of a given microstate m, so that [9]:

$$\wp_m = e^{-\beta E_m}/Q(T, V, N)$$

and;

$$\sum_m \wp_m = 1$$

This normalization, *i.e.* dividing the weighted probability of each microstate, $e^{-\beta E_m}$, by the partition function, is specifically required for counting the probabilities of the microstates in the Gibbs entropy, since these probabilities must, by definition, sum to one.

In relation to the macrostate of a system, the classical version of the canonical partition function is commonly adopted as a convenient way to establish the canonical ensemble (*i.e.* the set of all microstates that a closed system of specified volume, composition, and temperature can assume). In this approach, the microstate probabilities are computed in classical terms using the macroscopic properties of the system, *i.e.* they can be approximated as continuous variables. It can be useful to follow how the classical version of the canonical partition function is construed.

CHAPTER 9

Distributions of Particles and Energy

Abstract: The stability of a system is determined by the overall behaviour of the system's particles. In turn, this behaviour is established on the basis of the natural distributions the particles themselves spontaneously tend to assume. They tend to distribute across space according to a uniform spreading as the most probable outcome, and they also tend to share their energies according to a complex, non-uniform function that is nevertheless probabilistically equilibrated.

Keywords: Accessible states, Boltzmann factor, Classical Hamiltonian, Classical approximation, Configurational integral, Dominating configuration, Energy variance, Indistinguishable permutations, Levelling down energy, Macrostate fluctuation, Microstate probability, Minimum energy, Molecular partition function, Parcels equilibration, Sequence of microstates.

THE PHYSICAL ARRANGEMENT OF INDEPENDENT PARTICLES

To appreciate the probabilities of the distributions of particles and energy it is necessary to consider how energy distributes among particles and how particles actually distribute throughout space. The distribution of energy among particles occurs according to the Maxwell-Boltzmann equation, as we've seen. We are therefore going to see how particles actually distribute throughout space. To this end, it needs to be considered how the space over which particles distribute can be partitioned. Although the partitioning of energy is typically objective since the amount of energy of each particle has been long known to be quantized, that is, it can assume only a series of discrete values defined by Schrödinger's equation [9], the partition of particles over space is commonly approximated in a someway subjective manner and is mostly dependent upon the size of the particles (although, it will be mentioned, an objective quantization of space does seem to exist, and should then be a preferable framework for the calculation of the entropy).

It is first useful to consider the positional partition function, even called conformational partition function, in the case of macromolecules, or translational partition function, if we consider it as representing all the positions where the particles can translate to. This function is based on the partition of the system space into small units (typically, they should include only one particle each and therefore they should be of the size of the particles[26]). The positional partition function then considers the positional arrangement of the particles within the partition units in terms of combinatory calculus (as particle present/absent in each spatially localized partition unit, if the particles are all the same, or as permutations of empty/particle A/particle B/... particle Z, if there are more than one kind of particles[27]). In the most simplest approach, we can have a partition function for each given particle: the molecular (positional) partition function, q, which considers all the possible localizations of that particle throughout the system space[28]. If such particle is free to move throughout the system, the partition function for motion in three dimensions is $q = V/\Lambda^3$, where q is the total number of accessible positional states and Λ is the de Broglie thermal wavelength of the particles, which represents the space occupied by a particle[29] and which, incidentally, decreases with increasing particle mass and system temperature[30] [24]. For the whole ensemble of independent[31] and distinguishable N particles[32], the overall positional partition function[33] can initially be assumed to be just the product of the molecular partition functions of all the particles:

$$Q = q^N$$

which is the combination of the positional states of all the molecules[34]. However, if all the particles are identical and free to move through space, we cannot distinguish them (or, better, there is no objective reason to distinguish them) and therefore the relation $Q = q^N$ provides an overestimation of the possible distinct positional microstates (arrangements). In fact, it must be corrected for the number of indistinguishable positional microstates the system can assume[35] (which are indistinguishable because they are obtained by exchanging the positions of identical particles), thus, for indistinguishable independent atoms or molecules, the positional partition function of the ensemble can be better calculated as [24]:

$$Q = q^N/N! = (V/\Lambda^3)^N/N!$$

The nature of this partition function is evidently probabilistic[36]. It can be envisioned that even if we could, hypothetically, list and track the identical particles in the system so that we could keep a record at every subsequent instant of where each annotated particle is positioned, such a distinguishability would not be based on some objective distinction that systematically and persistently affects the individual properties of the particles; hence, all the microstates that appear identical though they are obtained by, for example, different permutations of the tracked particles, are indeed objectively identical. So, the $N!^{-1}$ correction is necessary because of the indistinguishable permutations that produce objectively identical microstates [17, 24]. It is not due to our incapacity to track particles. Things are very different if, instead, there is objective diversity among some particles that we are incapable of noting. An objective distinction does affect the probabilistic distribution of particles and, therefore, our incapacity of noting it would cause us to underestimate the value of the partition function, and, consequently, the real entropy.

NON-INDEPENDENT PARTICLES: THE CONFIGURATIONAL INTEGRAL

In reality, the positional partition function for the whole system, Q, is directly derivable from the products of the molecular partition functions of each of the single particles only if there are no interactions or joint effects, but this is typically an invalid assumption [25]. Thus, quite often, the statistical properties of the system are not simply the sum of the properties of all its constituent particles (*i.e.* the product of their molecular partition functions)[37], but they must be properly calculated as an average over all the microstates that the whole system can really assume[38] considering all its particles and their momenta together. The Q calculated in this way takes into account all the physical interactions occurring among the particles (such as electrostatic interactions, if the particles have a charge, or weak bonds like hydrogen bonds and van der Waals' interaction, if the particles are permanent or induced dipoles), as well as the effects the molecular partition functions can generate when they are considered jointly rather than separately[39]. This means that even the momentum (thermal) distribution of energy has now to be considered for the effects that can be generated when the particles are considered jointly rather than separately (always considering that, at a given T, the energy distribution must follow the Maxwell-Boltzmann equation, thus that the single particles cannot assume a whatever accessible momentum independently of the others). Likewise, the arrangements of the particles must ensure a uniform density distribution across the system volume. To this aim, the canonical partition function, wherein both the positional partition and the momentum (thermal) distribution of energy are introduced, for a real system can be calculated in classical (continuous) terms as:

$$Q(T, V, N) = 1/(h^{3N} \cdot N!) \cdot \int^{ps} e^{-\beta Em} \, dx_1 \, dy_1 \, dz_1 \, dEx_1 \, dEy_1 \ldots dEz_N$$

where the integration occurs over the whole phase space (ps; corresponding to $6N$ integrals for a system of spherical, homogeneous particles)[40] and h is Planck's constant, which provides a natural metric to make the canonical partition function dimensionless since it has units of [momentum] × [distance] [9]. Here, the summation over microstates is expressed by means of an integral because, for large systems, energy levels have huge degeneracies, and the spacings between neighbouring levels are nearly infinitesimal in comparison with the total internal energy [1]. In fact, at high temperatures quantum systems behave like classical ones because the energy spectrum becomes dense and appears continuous [9]. Together, the position of each particle and the momentum of the particle at each position correspond to the configuration of the microstate[41]. Thus, the configurational integral:

$$Z = \int^{ps} e^{-\beta Em} \, dx_1 \, dy_1 \, dz_1 \, dEx_1 \, dEy_1 \ldots dEz_N$$

is a probability density function of microstates over the phase space of the ensemble as it regards both the positional (spatial) distribution of the particles and their energies.

Notably, in the classical canonical partition function the focus is shifted from the hypothetical behaviour of independent particles to the thermodynamic properties the particles confer to a whole real system, with specific regard to the energy associated with the configuration (*i.e.*, microstate) of the system. That is, it looks at how the positional distribution of the particles in the volume of the system is linked to the energy of that configuration and then to the probability of the corresponding microstate, also considering the different levels of energy the particles can attain. In this way, the classical canonical partition function assigns decreasing probabilities to the product of the molecular partition functions whose joint effects cause an increase in the available energy of the system, or of parcels of the system as well (Fig. **3**). To appreciate the latter aspect, it can be noted that a parcel is any non-isolated part of a system and the different parts of a system at equilibrium must be balanced between each other, by definition, otherwise some unbalancing of forces is left, which would represent a source of potentially available energy. In fact, we can construct a system by adding together all its parts (in the sense of spatial parcels), and since the first law guarantees that the internal energy is additive, we can obtain the system's total internal energy as the sum of the contributions from all the pieces [1]. In other words, any macroscopic system can be considered as a combination of its parts as if they were smaller, more elementary systems. So, the configurational integral also applies to casual effects that move the state of the system away from the uniform distribution, or from a distribution equilibrated with external fields.

The casual effects, or fluctuations, depend upon the random perturbations of the distribution for either the particles over space (the position effect, for which the generation of a concentration/pressure gradient would correspond to a difference of energy between the parcels of the system that would be potentially available for doing work) or the kinetic energy among the particles (the momentum effect, for which any deviation from the Maxwell-Boltzmann distribution would generate a less probable microstate and thus would de-dissipate some heat, increasing the overall thermal energy of the system), or both (the interaction effect, for which the generation of potentially available energy would correspond to the generation of thermal gradients). So, any inhomogeneity in the internal energy of the parcels of a system could derive from a correlative synchronous behaviour of the particles (like a whirl of particles whose momenta should remain spatially associated for a discrete time, thus that some work could be theoretically "extracted" from the thrust of the moving fluid by, *e.g.*, a turbine or windmill), must be treated like a perturbation of the equilibrium and then as a matter of energetic equilibration between the parcels as established by their classical canonical partition functions. Indeed, if a macroscopic parcel behaves differently from the remaining system it can be considered as a subsystem with its own microstates and it is expected to equilibrate with the whole system, in the absence of specific constraints.

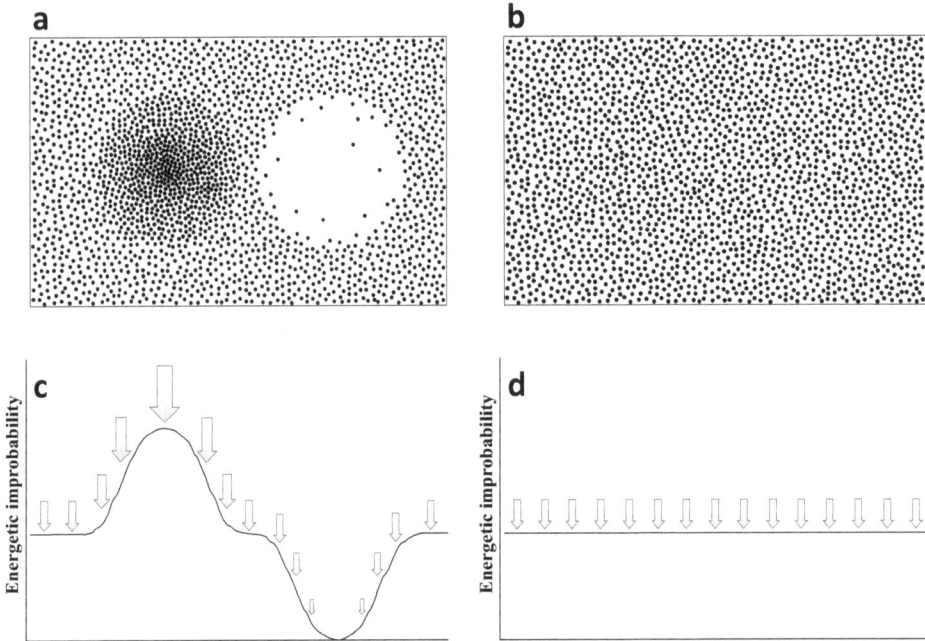

Fig. (3). The Boltzmann factor represents the weight of a given microstate in terms of probability. It assumes that the higher the internal energy of a microstate, the (exponentially) lower its probability of occurrence. The internal energy of a microstate depends upon many variables, like its concentration, thermal energy, intermolecular potential energy, and so on. In general, the improbability of a higher internal energy is considered to act as a levelling function that constrains the internal energy of a system to stay at a homogenous basal level throughout the system. Net transfers of energy and particles between systems, as well as between parcels of the same system, occur in consequence of gradients or discontinuities, which are picked up by the Boltzmann factor (which in this case is $e^{-\beta Em + \beta \mu Nm}$). A lower Boltzmann factor means a lower probability of the microstate of a system, or of a parcel of the system. At the macroscopic level this translates into an overall levelling function, since $\Omega(E^*, V, N^*)e^{-\beta E^* + \beta \mu N^*} \approx \sum_m e^{-\beta Em + \beta \mu Nm} = e^{-\beta PV/T}$. Thus, the probabilistic minimization of the fundamental thermodynamic potential that in the grand canonical ensemble is a transform of the overall entropy, and therefore a measurement of the equilibrium conditions, represents the overall macroscopic effect of minimization, at the microscopic level, of the full summation of probabilities through microstates (see text for explanation). For high energy levels, the Boltzmann factor tends to zero, that is, it declares these energies to have very low probabilities; whereas for energy levels tending to zero, the Boltzmann factor tends to one, *i.e.* 100% probability of occurrence for the corresponding microstate (but existing constraints typically fix the minimum accessible probability to a lower level, that is, each system is constrained to a given basal level of energy > 0, and then to an overall probability <100%). Anyway, the higher the energy and particles densities the stronger the probability they will be levelled down. In this way a general tendency to the most probable state is assured, which in most cases traduces as a function to spreading and sharing of matter and energy. In fact, (**A**) any discontinuity, or inhomogeneity, tends (**B**) to be levelled down because of the high improbability of any high-energy state of a system or of any parcel of a system. In this sense, the Boltzmann factor defines, in terms of energy, the diverse aspects of the physical world that tend to equilibrate in terms of probability, acting (**C**) like a sort of levelling function of energetic (im)probability (downward arrows). Entropy is the state function that measures the degree of levelling (in non-equilibrated systems) as well as the basal level of energetic probability of a system. Once the latter has equilibrated, (**D**) the system has reached its most probable state, and then the highest equilibration, stability, and inertness (as no work can be "extracted" from the system, unless some constraint is removed that makes a discontinuity available). The levelling function (downward arrows) is homogeneous throughout the system. It is worth noting that the levelling function of energetic (im)probability drives the system toward the most probable state, but that state is reached, or actualized, by means of the available energy of the system, which is thereby dissipated.

THE BOLTZMANN FACTOR: ENTROPY AS A MEASURE OF EQUILIBRATION, STABILITY, AND INERTNESS

As seen, the sequential microstates of a system are not independent; rather they are temporally connected since the macrostate fluctuates over microstates through time and every microstate is obtained from translations and collisions of the particles originating from the immediately antecedent microstate, and therefore these closely related microstates can be only minimally different (in positional and energy distributions). Thus, every microstate that corresponds to a macroscopic inhomogeneity of the temperature, or pressure, in the system (like one for which the higher energy particles are spatially separated from the lower energy ones, or one for which all the particles concentrate in half the space available) involves the existence of parcels with higher and parcels with lower energy. It is, however, highly improbable that some region spontaneously acquire a level of energy higher than the basic level [17]. In fact, the canonical partition function highlights that the microstates of higher energy have sharply decreasing probabilities and these inhomogeneities are all the more unlikely as they are more intense, as assumed by the Boltzmann factor (Fig. **3**). The particles have therefore the tendency to spread over the entire available volume because of their random motion, and expanding the volume per particle also increases the entropy. In fact, if all the particles are concentrated in half the space available as a consequence of a constraint, like a partition, removal of the constraint causes diffusion into previously unoccupied regions and approximately doubles the number of states accessible to each molecule (neglecting the volume physically occupied by the particles), and thus the number of states available to the system increases by a factor 2^N [1].

Indeed, any system can be considered as a combination of two (or more) subsystems or parcels, which can be spontaneously levelled only once any constraint that separated them has been removed. Hence, in the absence of constraints, every parcel should equilibrate its energy with the others (Fig. **3**) and whatsoever sequence of microstates that would tend to generate a gradient in temperature (or pressure) within a system becomes progressively much more likely to be reversed as it proceeds. This occurs because any relevant difference in temperature, or pressure, could be generated only through a sequence of microstates that progressively builds to a gradient of increasing intensity. However, this also would create an energy, thermal, or chemical gradient through the system[42]. The creation of such a gradient would correspond to generation of free energy (potentially available work in classical terms) and a lower number of microstates in statistical-mechanical terms, since it implies that some or all the particles would only access a reduced volume, and/or energy level, of the system [17]. Therefore, the maximization of Ω is just another facet of the same physical condition that maximizes the equilibration, stability, and inertness of a system[43]. It simply reflects the fact that, at equilibrium and in the absence of specific constraints, the macroscopic arrangement (state) of the system is such that its particles tend to occupy all the accessible physical and energetic microscopic arrangements because (i) the particles randomly move and exchange energy and (ii) the energy levelling function identified by the Boltzmann factor pushes them to arrange in a system (at each instant and then in each microstate) so that all and only the microscopic arrangements that minimize the free energy of the system (or, in general, the parameter that in each diverse condition has to be minimized for having the maximum stability, equilibrium, and inertness of the system) are visited and, specifically, are visited with equal frequency. The occupancy of all the accessible microstates is then the necessary result of the particles moving and exchanging energy under the permanent pressure of the probabilistic levelling function identified by the Boltzmann factor, which is just a measure of the net force, derived from the motions of the particles, that pushes the system to the equilibrium. The fluctuation of a given macrostate over all the accessible microstates[44] of the phase space is therefore a sufficient and necessary condition that assures that there is no available energy in the system and therefore the latter is equilibrated, stable, and inert. This is why macrostate probability and the number of microstates are equivalent expressions of the same state function, *i.e.* entropy. In a dynamic equilibrium, fluctuations involve local transient decreases of entropy and tiny imbalance of forces. The latter tends to restore the entropy back to its maximum, thus the fluctuations around equilibrium are random and ephemeral. So, relevant inhomogeneities in the system cannot be spontaneously generated, even transitorily, by random fluctuations of the macrostate over microstates, except with the utmost improbability; actually they are never observable under everyday conditions. In fact, the second law prescribes that an isolated system at equilibrium will maximize its entropy, and as a consequence, at constant volume, particles and entropy (that is, exactly at equilibrium), E is minimized with respect to any internal perturbations of the system [9]. In other words, since any transient increase in the density, or energy, in some spatial region is a perturbation away from equilibrium, it follows that, throughout an isolated system at equilibrium, the internal energy is maintained at its minimum, *i.e.* it is levelled down to its basal level, as shown in Fig. (**3**). This is also known as the principle of minimum energy, and it is just another interpretation of the second law of thermodynamics [9].

It should be noted that, for an ideal gas, ephemeral spots of more concentrated particles (Fig. **3**), which could be casually generated by the random positional arrangement of the particles, would occur through a flow of particles against a gradient of increasing concentration, that is, towards a higher pressure, or chemical potential, and then a higher overall available energy would be spontaneously generated. In this instance, no temperature change would occur to compensate for the increasing energy because it is only a difference in the intermolecular potential energy that affects the mobility of the particles and therefore their kinetic energy, whereas by itself the concentration component of the chemical potential does not affect the temperature [4]. In fact, it can be seen as a free compression, wherein the kinetic energies of the particles transfer from each other, and then no net change in the overall kinetic energy of the self-compressing gas parcel occurs. The case would be the same for the casual generation of regions with higher concentration of a given solute in a liquid: the simple concentration effect would increase the chemical potential and therefore the internal energy of a parcel at the expense of some other parcel. No compensation with a corresponding change in the temperature can occur for an ideal solution as a consequence of a change in the solute concentration. However, in many real fluids, there can be an effect on the temperature, due to changes in the intermolecular potential energy.

Clearly, eventual constraints limit the full accessibility to all the otherwise possible microstates; therefore, they reduce entropy. For example, a gravitational field, by generating a vertical gradient of gas concentration (actually appreciable only if the gaseous system is much extended vertically), causes a reduction in the overall number of accessible microstates. In the absence of specific constraints (*e.g.*, external fields), the canonical partition function assumes a macroscopically uniform density of particles throughout the system volume, the Maxwell-Boltzmann distribution for thermal states among particles, and a random spatial distribution of thermal energy levels, as representing the maximally probable distribution of particles on a $6N$ phase space. Thus, for the canonical ensemble of a closed system the probability $\wp_m$ that the system occupies a microstate m is $\wp_m = e^{-\beta E_m}/Q(T, V, N)$, that is, the probability of the occurrence of a given microstate is an exponentially decreasing function of its internal energy E_m (since $k_B T$ is the thermal component of the internal energy, typically, for real systems, $E_m > k_B T$ and therefore $\beta E_m > 1$).

THE DOMINATING CONFIGURATION: FLUCTUATIONS AND THE BOLTZMANN FACTOR

Microstates of a closed system equilibrated with its surroundings, or of equilibrated parcels inside a system, that move away from this condition, *i.e.* whose internal energy shows casual fluctuations toward higher energy levels (of the whole system with respect to its surroundings or in some parcel of the isolated system, in which case the energy must be lower in other parcels, as the system's energy cannot change), are therefore very improbable, and, in fact, any energetic fluctuation is most commonly tiny and always transient. Among all the theoretically accessible microstates, the subset that has an internal energy (essentially a thermal energy, since this is the one that can fluctuate between the system and its surroundings) close to the minimum possible (given the fixed temperature of the thermal reservoir) is therefore the one that is largely most probable. This subset is called the dominating configuration [25]. Indeed, for a system that can assume a very large number of microstates, there is typically a dominating configuration that is overwhelmingly the most probable and dominates the properties of the system virtually completely, that is, we can evaluate the thermodynamic properties of the system by taking that single as representative of the average over the whole ensemble [25]. In fact, the fluctuations of the system outside the dominating configuration have a sharply decreasing probability as the energy distribution departs far from it. We will see that the same principle is followed in other ensembles that exhibit particle transfer.

If it is assumed that no interactions occur between the position and the thermal energy of the particles (*i.e.*, the temperature is uniform throughout the system, which in large real systems at equilibrium may be a reasonable assumption since microstates implying inhomogeneities in the temperature are highly improbable in the canonical ensemble)[45] then the microstate energy is given by the so-called classical Hamiltonian, that is, it is the sum of potential and kinetic energies [9]:

$$E_m = H(P^N, r^N) = K(P^N) + U(r^N)$$

where P and r represent the momentum and position of each of the N particles, respectively, and where $H(P^N, r^N)$ is a classical Hamiltonian function that expresses the internal energy E of the system, $K(P^N)$ is a kinetic energy function depending on the momenta of the particles, and $U(r^N)$ is a classical potential energy function that depends on the positions but not the momenta of all of the particles[46] [9]. This approach allows us to separate the kinetic and positional energy contributions to the canonical partition function. In addition, because in the canonical ensemble the temperature is fixed, the partition of thermal energy over particles may be considered practically fixed as well, and can then be removed from the sum and included in the calculations as an overall constant. It therefore becomes possible to simplify the classical partition function further [9], so that:

$$Q = Z/(\Lambda^{3N} \cdot N!) \text{ with } Z = \int^V e^{-\beta Em} \, dx_1 \, dy_1 \, dz_1 \, \dots \, dz_N$$

or,

$$Q = \int^V e^{-\beta Em} \, d\tau_1 \, d\tau_2 \, \dots \, d\tau_N$$

wherein $d\tau_1$ is shorthand for $dx_1 \, dy_1 \, dz_1$ with the integration occurring over the volume of the system[47] for each microstate. In this way, the kinetic contribution can be evaluated analytically and, for simple spherical particles, it gives the thermal de Broglie wavelength [9], as already preliminary noted for the molecular partition function. What is left in the classical partition function, as probabilistically dependent upon the Boltzmann factor, is just the energy derived from the positional and intermolecular effects of the particles. Thus, the positional energy contribution gives rise to a configurational partition function for the canonical ensemble [9]. As seen, over the volume of a large system, the energy can be macroscopically considered to be distributed as a continuous variable, and this is why the configurational partition function of the ensemble includes an integral.

Finally, the Boltzmann factor includes β in the exponential term that defines the probabilities of the microstates ($-\beta E_m = -E_m/(k_B T)$) because the probability of a microstate does not depend only and negatively on the intensity of the eventual available energy (generated by the removal of a constraint or by random fluctuations/perturbations) that it can involve, but also, positively, on the average thermal energy of the particles. In fact, the higher the temperature the wider the fluctuations and then the higher the probability that a time sequence of microstates can proceed toward a given E_m and, therefore, the improbability of such event is proportional to $E_m/(k_B T)$. This is owing to the fact that higher temperatures correspond to a more spread distribution of thermal energy throughout energy levels, which means larger differences in energy among particles. Therefore, configurational inhomogeneities have greater probability of occurring in stronger randomly-generated perturbations. Thus, because $E_m \propto T$ (since, for an ideal system, $E = E_0 + c_V T$), assuming $-E_m/(k_B T)$ in the exponential term provides a correction for the increment in the improbability of E_m that is due to T. In fact, in macroscopic systems at constant temperature, fluctuations in energy are very small, and can be specifically quantified by calculating the energy variance[48] [9]:

$$\sigma_E^2 = c_V k_B T^2$$

This means that the magnitude of energy fluctuations, which anyway are far too small to detect macroscopically, is actually related to the constant volume heat capacity and to the temperature squared. In addition, from this relationship it can be shown that, although the absolute fluctuations increase with the system size because $c_V \propto N$, the relative magnitude of energy fluctuations decreases as the inverse square root of the system size [9]. In fact, relative to the average, these fluctuations decay as $N^{-1/2}$ so that macroscopic systems appear to have constant, well-defined energies.

CHAPTER 10

Boltzmann Entropy & Equilibrium in Non-Isolated Systems

Abstract: The microscopic approach of statistical mechanics has developed a series of formal expressions that, depending on the different features of the system and/or process involved, allow for calculating the value of entropy from the microscopic state of the system. This value is maximal when the particles attain the most probable distribution through space and the most equilibrated sharing of energy between them. At the macroscopic level, this means that the system is at equilibrium, a stable condition wherein no net statistical force emerges from the overall behaviour of the particles. If no force is available then no work can be done and the system is inert. This provides the bridge between the probabilistic equilibration that occurs at the microscopic level and the classical observation that, at a macroscopic level, a system is at equilibrium when no work can be done by it.

Keywords: Approximate equiprobability, Approximate isoenergeticity, Boltzmann entropy, Boltzmann factor, Canonical ensemble, Canonical partition function, Dominating configuration, Energetic (im)probability, Equal probabilities, Equilibrium fluctuations, Fundamental thermodynamic potential, Gibbs free energy, Grand canonical ensemble, Helmholtz free energy, Maximization of entropy, Microcanonical partition function, Microcanonical system, Minimization of energy, Temperature, Thermostatic bath.

RELATIONSHIP BETWEEN THE MICROCANONICAL (ISOENERGETIC) AND CANONICAL (ISOTHERMAL) ENSEMBLES

The classical Ω is calculated, for a system at equilibrium defined by given N, E, and V (*i.e.* a microcanonical system), by counting, among all the possible micro-

Alberto Gianinetti

states (*i.e.* every possible combination of positions, orientations, and velocities for all of the particles), only the number of microstates whose energy equals E, that is, those that belong to the microcanonical ensemble [9]. In fact, in the microcanonical ensemble, there are absolutely no fluctuations in energy, since it assumes that N, E, and V are constant, so that, by hypothesis, no fluctuations of E can occur over microstates. These microstates are assumed to be equiprobable, and $\Omega(E, V, N)$ represents the microcanonical partition function associated with such a condition [9]. In most real systems, however, the assumption that no fluctuations of E can occur over microstates cannot be guaranteed, because they are not isolated. This is why, in practice, the canonical partition/ensemble, which assumes that N, T, and V, are constant for a system, appears more realistic, since it is quite easy to guarantee the stability of the system temperature by, for example, a large thermostatic bath. In the canonical ensemble the probability of each microstate m is proportional to $\exp(-\beta E_m)$, where E_m is the energy of that microstate, that is, in the canonical ensemble the system visits each microstate with a frequency proportional to its Boltzmann factor [9]. The energy fluctuates in the canonical ensemble, even if only by very small amounts, while it is the temperature that is kept constant by coupling with an infinite heat bath [9]. Note, in fact, that, whereas the internal energy of a system, E, is an exact quantity (whatever the size of the system is) that can be theoretically kept perfectly constant if the system is perfectly insulated, the system's temperature, T, is an overall property that emerges from the statistical average behaviours of all the particles. So, temperature becomes increasingly reliable as a macroscopic property as the number of particles that contribute to determine it gets larger and relative deviations from this average value get smaller. Ultimately, temperature can be considered an exact property only in the limit of an infinite number of particles, that is, in the presence of an infinite thermal reservoir (the heat bath) with which the system (even a small one) is thermally equilibrated.

The fact that, in actuality, the microstates of every non-isolated system are not necessarily all equiprobable would suggest that the original Boltzmann equation, which assumes Ω as the number of microstates the system can access at equilibrium, provides a biased estimation of the maximum entropy outside microcanonical conditions. Specifically, it would seem to underestimate its value since there are many possible (although very improbable) microstates that are excluded [9] because they have higher energies than the one theoretically assumed for the microcanonical ensemble of the system. Indeed, the Boltzmann equation requires that entropy is calculated as $S(E, V, N)$ for an isolated system at equilibrium, that is, the energy, the volume, and the number of particles are exactly defined. However, E can be truly considered constant only if the insulation of the system is really perfect, which is a very difficult task to accomplish. On the other hand, real systems can easily be kept at a given temperature by means of a thermostatic bath, that is, it is easy to keep them in canonical conditions, wherein T, V, and N are fixed, but it is not really possible to guarantee that every microstate of a system has a given, exact energy level, as the microcanonical assumptions would require, since microscopic fluctuations in the system (resulting from microscopic transfers of thermal energy between the system and the bath, or the surroundings, which are in dynamic thermal equilibrium) can alter, even if passingly, this condition. In other words, strictly speaking, microcanonical conditions would seem quite imaginary for most ordinary systems. Nonetheless, it is important from a theoretical point of view to consider entropy under microcanonical assumptions, as $S(E, V, N)$ is a fundamental thermodynamic potential (*i.e.*, all the thermodynamic properties of the system can be derived from it, specifically in terms of its derivative, given the values of the fixed state functions), whereas $S(T, V, N)$ is not [9]. This means that the microcanonical entropy is a much more meaningful parameter than the canonical one, and, for this reason, it should be preferentially adopted for thermodynamic interpretations when possible.

An excellent compromise to these contrasting exigencies (have E theoretically fixed, so that entropy has a deep thermodynamic value, and to fixing the temperature as an actual possibility) is actually obtained thanks to the fact that the peak in the distribution of probabilities of energy levels over the microstates of the canonical ensemble (wherein T, instead of E, is fixed) is so sharp that the probability of energy $\wp(E)$ appears essentially to be one for only a single energy, and zero for all others [9], that is, in the canonical ensemble the system fluctuations occur almost entirely over microstates of energy E^*, the level of internal energy that is determined by the temperature of the system. In other words, as the dominating configuration of the canonical ensemble is constrained to energy E^* by the fixed system temperature T (guaranteed by the theoretically infinite thermal reservoir), then the ensemble identified by $Q(T, V, N)$ is actually constrained to fluctuate essentially only upon $\Omega(E^*, V, N)$ microstates. Thus, the energy E^* represents the almost unique possible energy level of the microstates of the canonical ensemble just because $Q(T, V, N)$ is constrained to T. Hence, all the actual microstates of the canonical ensemble can essentially be considered equiprobable and the original Boltzmann equation holds with good approximation even in such case. Therefore, the condition of fixed temperature for the canonical ensemble, in practice, approximates the condition of fixed energy for the microcanonical ensemble, at least for large systems far from absolute zero. Thus, if we approximate the canonical ensemble as one in which only the most probable energy E^* is visited, we may estimate Q according to the equivalence [9]:

$$Q(T, V, N) \approx \Omega(E^*, V, N) \cdot e^{-\beta E^*}$$

that is, the canonical ensemble is also approximately isoenergetic, as it can be adequately represented by the sole microstates of the dominating configuration of energy E^*. For this reason, the microcanonical Ω can indeed be considered a valid approximation for the number of microstates in the dominating configuration of the canonical ensemble and then of the canonical ensemble itself. The actual number of practically accessible microstates of the canonical ensemble is therefore virtually the same as that of the microcanonical ensemble[49] [17]. The two partition functions (*i.e.* the one for the canonical ensemble and the one for the microcanonical ensemble) are however numerically different because the canonical one is weighted by the Boltzmann factor for the system's energy, whereas in the microcanonical partition every microstate is fully counted if it has the fixed level of energy or is not counted at all. Thus, in the canonical partition function the Ω microstates of energy E^* are weighted by their energy, so that [9]:

$$\wp(E^*) = \Omega(E^*, V, N)e^{-\beta E^*}/Q(T, V, N) \approx 1$$

This relationship expresses the probability of the dominant configuration of the canonical ensemble, to which the $\Omega(E^*, V, N)$ microstates whose internal energy is centred to E^* belong, and which, therefore, have a probability of $e^{-\beta E^*}$. They stay at that level of probability because, by hypothesis, they are constrained to temperature T, and then to energy E^*, by the thermal reservoir.

In addition, though the classical $\Omega(E, V, N)$ should be calculated by counting among all the possible microstates only the number of those microstates whose energy equals E [9](which is actually quite difficult to calculate), the classical canonical partition function of the system at equilibrium, $Q(T, V, N)$, can be used to approximate the value of Ω. The trick is to avoid the complicated issues of combinatorics by evaluating sums over microstates [9], that is, by considering the macroscopic properties of the system, as previously remarked for the classical version of the canonical partition function. Actually, the systems that are dealt with in thermodynamics contain a huge number of particles, and therefore it would be practically impossible to specify all the $6N$ variables (or more, in the case of real molecules) required by the statistical approach based on the microscopic mechanics of the system. A detailed knowledge of the position and motion of each particle would even be superfluous since thermodynamics is rather concerned with the average properties of the system [6]. In this regard, an outcome that turns out to be very favourable is that the Boltzmann equation provides an estimation of the maximum entropy the system can attain, equal to that provided by the Gibbs equation applied to the equilibrium [9]. This is because a real system at temperature T actually fluctuates over microstates over the dominating configuration having approximately equal probability, and eventually also over microstates outside of the dominating configuration, but with a probability that is approximately equal to zero.

THE ROLE OF THE HELMHOLTZ FREE ENERGY IN CANONICAL (ISOTHERMAL) CONDITIONS

Although the microcanonical partition function $\Omega(E, V, N)$ maintains its role in calculating entropy even in canonical conditions because:

$$S(T, V, N) \approx S(E^*, V, N) = k_{\mathrm{B}} \ln \Omega(E^*, V, N),$$

in these conditions its role as a fundamental thermodynamic potential, specifically aimed at determining equilibrium conditions, is taken up by its homologous expression, that is, the logarithm of the canonical partition function, which is a fundamental thermodynamic potential in isothermal conditions [9, 14]:

$$A(T, V, N) = -k_B T \ln Q(T, V, N)$$

where the factor $k_B T$ makes clear that $A(T, V, N)$ is a kind of energy. Specifically $A(T, V, N)$ is the Helmholtz free energy, which for closed systems held at a constant temperature and volume, is the state function that reaches a minimum at equilibrium [9]. Actually, this function is minimized when the thermal energy, and therefore T, is uniform and constant [1]. That is, it guarantees that, at equilibrium, the temperature stays uniform throughout a system or over thermally connected systems. In general, the logarithm of the partition function for any ensemble, multiplied by k_B, gives a macroscopic thermodynamic potential that is a transform of the entropy [9], and is therefore specifically aimed at determining equilibrium conditions. The transform is performed with respect to the fluctuating macroscopic variables in the ensemble, and therefore it represents the appropriate state function to define the system's equilibrium, because the system's entropy is not a fundamental thermodynamic potential outside isoenergetic systems [9]. This is because, outside those systems, it is not the system's entropy but the overall entropy that determines the equilibrium state of the system.

From the above definition of the Helmholtz free energy, it can be derived that the value of the canonical partition function of a system at equilibrium equals the value of a Boltzmann factor that weights the energetic (im)probability for the whole system according to its Helmholtz free energy:

$$Q(T, V, N) = e^{-\beta A}$$

and since the canonical ensemble can be approximated as one in which the sole microstates of the dominating configuration of energy E^* are visited, that is:

$$Q(T, V, N) \approx \Omega(E^*, V, N) \cdot e^{-\beta E^*}$$

it turns out that;

$$e^{-\beta A} \approx \Omega(E^*, V, N) \cdot e^{-\beta E^*}$$

Thus,

$$\ln e^{-\beta A} \approx \ln \Omega(E^*, V, N) + \ln e^{-\beta E^*}$$

and then the Helmholtz free energy is shown to be linked to other state functions of a closed system:

$$A(T, V, N) \approx E^* - k_B T \ln \Omega(E^*, V, N) = E^* - T \cdot S(E^*, V, N)$$

In addition, as

$$E(T, V, N) \approx E^*$$

then;

$$A(T, V, N) = E(T, V, N) - T \cdot S(T, V, N)$$

at least when fluctuations are neglected. So, under isothermal conditions:

$$A = E\text{-}TS, \text{ or } E = A + TS$$

that is, the internal energy of a closed system can be partitioned between a freely available component, the Helmholtz free energy, and a non-free component, TS, with the former being available to do some work (this depends upon its having an intrinsic improbability, identified by $e^{-\beta A}$), while the other is the equilibrated, stable, and inert energy component, as established by the system's entropy.

This is quite interestingly, since:

$$Q(T, V, N) \equiv \sum_m e^{-\beta E_m}$$

then, based on the equivalence found above,

$$e^{-\beta A} = \sum_m e^{-\beta E_m}$$

that is, at a macroscopic level, the value of a Boltzmann factor that weights the energetic (im)probability for the whole system according to its Helmholtz free energy is equivalent to the sum of Boltzmann factors that weight the energetic (im)probability for each microstate of the canonical ensemble.

As the dominating configuration can be used as a representative of the whole ensemble and therefore the Boltzmann factor of the dominating configuration multiplied by its number of microstates can be used in place of the full summation through microstates, we also have:

$$e^{-\beta A} \approx \Omega(E^*, V, N) \cdot e^{-\beta E^*}$$

In fact, the canonical partition function is so sharply peaked that the partition sum of individual microstates can be approximated with a single value of the summand, the value that gives the most likely state of the system, ignoring small fluctuations [9]. Since the latter summation is minimized when all the microstates have essentially the same energy, that is, under almost isoenergetic conditions, then the Helmholtz free energy is also minimized under the same conditions. In other words, the free energy of the system acts as an overall levelling function. As remarked by Lee [14], this equation may therefore be regarded as connecting the microscopic world that we specify with microstates and the macroscopic world that we characterize in terms of free energy. Whereas the entropy is the fundamental thermodynamic potential that is maximized in isolated systems to attain the equilibrium, in systems that are in thermal contact with a reservoir, or surroundings, at temperature T, the equilibrium is attained when the system's fundamental thermodynamic potential A(T, V, N) is minimized. So, for systems interacting isothermally, the second law demands that changes in the Helmholtz free energy must be negative, reaching a minimum at equilibrium [1].

The rationale of the above finding is that, in a closed system, it is not the system entropy that is maximized, like in an isolated system, but, rather, the overall entropy of the system plus that of its surroundings, since the two interact. This is precisely what the Helmholtz free energy looks at; in fact, as A = E-TS, a change in A is calculated as:

$$dA = dE - TdS$$

and since any change in the internal energy is actually a transfer of heat, then:

$$dA = \delta Q - TdS, \text{ and } dA/T = \delta Q/T - dS$$

wherein by S, it is specifically meant S_{sys} as A is a state function of the system. Thus, a transfer of heat to the system determines the increase of entropy in the system, but, at isothermal conditions, if the system and the thermal reservoir are in thermal equilibrium then only infinitesimal heat transfers can occur. The Clausius' equation then applies, thus that $\delta Q/T = dS$, that is, they are equal and therefore there is no change in the Helmholtz free energy of the system. So, at equilibrium, in any process that involves an isothermal transfer of heat from/to the environment, there is a change of entropy in the system that is equal and opposite, thus that there is no change in the Helmholtz free energy of the system. In fact, at equilibrium $dA = 0$ and then $dE = TdS$ thus that $dS = dE/T$. Nevertheless, a discrete transfer of heat occurs for spontaneous processes, and in such case $\Delta A < 0$; in fact, the free energy decreases in any process that approaches the equilibrium. Thence, on the one hand, at thermal equilibrium, a transfer of heat corresponds to a change in entropy of the surroundings, that is:

$$\Delta E/T = -\Delta S_{sur}$$

and then, at equilibrium:

$$\Delta S_{sys} = -\Delta S_{sur}$$

On the other hand, a process is spontaneous if the total entropy increases and then if $\Delta A < 0$; in fact, if $\Delta S_{tot} = \Delta S_{sys} + \Delta S_{sur} > 0$, then for a spontaneous process:

$$\Delta S_{tot} = \Delta S_{sys} - \Delta E/T > 0$$

that equals to;

$$T\Delta S_{tot} = T\Delta S_{sys} - \Delta E > 0$$

and then to;

$$-T\Delta S_{tot} = \Delta E - T\Delta S_{sys} < 0$$

from which we have;

$$-T\Delta S_{tot} = \Delta A < 0$$

thus it can be finally concluded that;

$$\Delta A < 0$$

In other words, $-\Delta A$ is the energy isothermically dissipated in any spontaneous process that occurs at canonical conditions. It should be noted that, for these formulas can be applied, the process needs to be isothermal (like discussed for the

Clausius' equation), that is, at the temperature of the thermal reservoir, but since a discrete transfer of heat is hypothesised to occur, some process that alters the thermal equilibrium between the system and its surroundings must take place. As previously seen, the fact that $-\Delta A$ is dissipated in an equilibration process, *i.e.* it is no longer free, is based on the fact that a process is spontaneous and irreversible if the overall entropy increases. There is therefore a perfect equivalence between the maximization of the overall entropy and the minimization of the free energy (the Helmholtz one, in this case) of the system.

THE GIBBS FREE ENERGY IS USED IN MANY EVERYDAY (ISOBARIC) CONDITIONS

In transfers of energy under canonical assumptions, the system's volume is fixed by hypothesis; however, in many everyday processes, there is no constraint on volume changes, but, rather, they happen at atmospheric pressure, meaning that they are isobaric. For example, water freezes at 0 °C, but part of the energy released during this change of state is used by the forming ice to expand in volume against atmospheric pressure. Under these conditions another kind of free energy, the Gibbs free energy (which is widely used in chemistry), is the proper state function used to calculate the equilibration of the system. It is obtained by adding the *PV* work term into the expression of the Helmholtz free energy:

$$G(T, P, N) = E + PV\text{-}TS = \text{H - } TS$$

Again, the last term, $-TS$, indicates that the higher the entropy of the system the more its energy is dissipated, *i.e.* it is no longer available for doing work. Indeed, an interpretation of the Gibbs free energy is that it is the difference between the total energy and the energy stored chaotically [29]. The Gibbs free energy is thus the energy available in the form of utilizable work for closed systems at constant pressure and temperature. H $= E + PV$ is the enthalpy, that is, the sum of the system's internal energy plus the product of P and V of the system. Thus, enthalpy includes the change of internal energy, as well as any PV work done on the system, that occurs in a given process. In fact, if the process is both isothermal and non-diffusive ($dT = dN = 0$), the Gibbs free energy measures the work done by the system as $dG = -PdV$. In these conditions, the Gibbs free energy is minimized (*i.e.*, it is most negative) when at a given pressure the volume of the system is maximized; in fact, a gaseous system tends to expand against a given external pressure until its internal pressure equals the external one. In these conditions, the Gibbs free energy measures the PV work that can be done by the system. In general, the Gibbs free energy is minimized when both T and P are uniform and constant, that is, it guarantees that, at equilibrium, temperature and pressure stay uniform throughout a system or over thermally and mechanically connected systems.

Like the Helmholtz free energy, the Gibbs free energy, G, embodies the concept that a process is irreversible if the overall entropy increases, and this is why it is used to evaluate the spontaneity of chemical processes, as:

$$\Delta G = \Delta H - T\Delta S$$

where S specifically means S_{sys} since, like A, even G is a state function of the system. Chemical reactions are spontaneous if $\Delta G < 0$, but, notably, $-\Delta G$ is plainly the quantity of energy that can be dispersed to the surroundings in a given chemical reaction or process [30], as it is merely a derivation of the overall entropy change [31]. In fact, analogously to the Helmholtz free energy, at equilibrium $\Delta G = 0$ and thus $\Delta H = T\Delta S$, then $\Delta S = \Delta H/T$. In plain words, at equilibrium, the change of entropy in the system due to the reaction equals the amount of heat that is isobarically produced by the reaction; this heat is then isothermically transferred to the environment. By definition, such heat transfer corresponds to the change of entropy of the environment, that is:

$$\Delta H/T = -\Delta S_{sur}$$

and then, at equilibrium

$$\Delta S_{sys} = -\Delta S_{sur}$$

On the other hand, a chemical reaction will take place, *i.e.* it is spontaneous, if $\Delta G < 0$ and then if the total entropy increases; in fact, if:

$$\Delta S_{tot} = \Delta S_{sys} + \Delta S_{sur} > 0$$

then, under isothermal and isobaric conditions, for a spontaneous irreversible reaction:

$$\Delta S_{tot} = \Delta S_{sys} - \Delta H / T > 0$$

that equals to;

$$T\Delta S_{tot} = T\Delta S_{sys} - \Delta H > 0$$

and then to;

$$-T\Delta S_{tot} = \Delta H - T\Delta S_{sys} < 0$$

from which we have;

$$-T\Delta S_{tot} = \Delta G < 0$$

thus it can be finally concluded that for the spontaneous process;

$$\Delta G < 0$$

In other words, $-\Delta G$ is the energy isothermically dissipated in an isobaric process [4]. As remarked by Strong and Halliwell [15] a reaction proceeds when it leads to an increase in total entropy and the reaction stops when no way is open for a further increase of total entropy. Reactions that occur at atmospheric pressure will then stop when the Gibbs free energy is minimized.

OPEN SYSTEMS

Last, we can envision that within every system, or between open systems, there can be a movement of both particles and energy. In fact, whereas in canonical conditions the Boltzmann factor considers the energy of the system as the only variable that determines the levelling function, when particles can be exchanged, like in an open system, the chemical potential must be considered beside the thermal energy, and therefore the Boltzmann factor becomes $\exp(-\beta E_m + \beta\mu N_m)$, that is, the chemical potential, which we shall consider later, enters into the levelling function to determine the equilibration of the particles in the so called grand canonical ensemble, whose partition function is $\Xi(T, V, \mu)$ [9]. This ensemble is particularly interesting when considering the distribution of matter and energy throughout a fluid system, where the equilibration occurs between each macroscopic parcel and the rest of the system (Fig. **3**). In general, in any ensemble, the joint probability distribution for the fluctuating macroscopic state functions is related to the microscopic Boltzmann factor multiplied by the microcanonical partition function Ω [9]. In this way, just like in the canonical ensemble, the macroscopic distributions of the fluctuating macroscopic state functions in every ensemble are sharply peaked, for large systems. Thus, these fluctuations are too small to observe in macroscopic systems [9]. Hence, in the grand canonical ensemble, the probability of the dominant joint distribution of energies and particle numbers is:

$$\wp(E^*, N^*) = \Omega(E^*, V, N^*)e^{-\beta E^* + \beta \mu N^*}/\Xi(T, V, \mu) \approx 1$$

Notably, the fundamental thermodynamic potential that is minimized at equilibrium conditions in the grand canonical ensemble is [9]:

$$PV/T$$

As with any ensemble, the logarithm of the partition function multiplied by k_B, gives a macroscopic thermodynamic potential that is a transform of the entropy [9] and is specifically aimed at determining equilibrium conditions. As it was shown for the Helmholtz free energy, it can be derived that:

$$e^{-\beta PV/T} = \sum_m e^{-\beta Em + \beta \mu Nm} \approx \Omega(E^*, V, N^*)e^{-\beta E^* + \beta \mu N^*}$$

Since the summation over microstates is maximized when all the microstates have essentially the same energy and particle densities (*i.e.*, at equilibrium), then PV/T is minimized under the same conditions (actually, it becomes defined at the minimum possible value, because we cannot even numerically define the temperature, pressure, or chemical potential for a system that is not in equilibrium [1]). In other words, PV/T acts as an overall levelling function (between parcels of the same system or between connected systems), and, also in the case of the grand canonical ensemble, the Boltzmann factor provides a close connection between the microscopic and the macroscopic worlds. Since T and V are fixed by hypothesis, it is then P that is minimized, and, accordingly, equalized throughout the system. This can be intuitive, because P increases by both increasing E (and thus the kinetic energy of the particles of a parcel at the expense of the rest of the system; although fluctuations in E can only be minimal once T is fixed) and N, and therefore minimizing them both equates to minimizing P. For an ideal gas:

$$PV = \mathrm{n}RT = Nk_{\mathrm{B}}T$$

and then

$$PV/T = k_{\mathrm{B}}N$$

Hence, PV/T has the same measure unit as k_{B} (and then as entropy), that is, J K^{-1}, and therefore it represents the amount of available energy per Kelvin degree (thus, it is the free energy per temperature increment that is available for doing work based on particle transfer). For an ideal gas it merely depends on the number of particles in the system, that is:

$$P \propto N$$

This aspect is what makes the distribution of particles in Fig. (**3**) intuitive: the number of particles for a given volume (intended as number of spots for given area, in a bi-dimensional representation) must be uniform for the system to be equilibrated with respect to the matter distribution. The full role of P in smoothing any macroscopic fluctuation of both particle concentration and energy (and thereby its usefulness in assessing the equilibration of the system) could be rendered in a colour figure by assigning different colours to particles having different kinetic energies and showing that even the colour distribution is essentially uniform throughout the system. For systems where more species of particles are present the partial pressure of each species must be constant throughout the system. This holds even for a liquid at equilibrium, since, as we will see, the chemical potential of each species in the liquid is in equilibrium with its partial pressure in the gaseous phase. Hence, at equilibrium, the overall pressure tends to be strictly uniform throughout every system, be it homogeneous or heterogeneous (*i.e.*, including different phases). In fact, even liquid and solid phases always equilibrate to the pressure of the gaseous phase, dependent on the temperature and the substance. If the system includes different substances, each of them corresponds to a partial pressure that is equilibrated throughout the system; the overall pressure is equilibrated accordingly [9, 25, 32].

GENERAL RELATIONSHIPS BETWEEN THE MACROSCOPIC AND THE MICROSCOPIC PROPERTIES OF THE SYSTEM

The three expressions equated above provide equivalent measures of the equilibrium in an open system based on the Boltzmann factor and therefore they are all maximized at equilibrium (this means that the energetic probability of the system is maximized and thereby its improbability is minimized; (Fig. 3)). However, they characterize the equilibrium at different levels. The first expression:

$$e^{-\beta PV/T},$$

demonstrates that, at the macroscopic level, the most probable state, *i.e.* the equilibrium, occurs when *PV/T* is minimized, that is, the capability of the system to do work, as measured by the proper thermodynamic potential, is minimized (under the hypothesized conditions and constraints); the second expression:

$$\sum_m e^{-\beta Em + \beta \mu Nm}$$

indicates that, at the microscopic level, the equilibrium condition implies that the sum of the microstates probabilities is maximized (in fact, when *PV/T* is minimized $e^{-\beta PV/T}$ is maximized and so is $\sum_m e^{-\beta Em + \beta \mu Nm}$). This occurs when all the microstates that belong to the predominating configuration can really be accessed, whereas microstates with greater energy (either thermal or chemical), and therefore lower probability, are not actually accessed (although they could be accessed because of ephemeral, random fluctuations of matter and energy between the open system, or parcel, and its surroundings). In fact, such sum is maximized only when the microstates actually visited by the system are essentially restricted to those having the minimal energy and number of particles that the constraints to a given temperature and chemical potential can admit, that is, the microstates that belong to the predominating configuration. Of course, ephemeral, random fluctuations of matter and energy could also cause the occurrence of microstates with energy lower than that belonging to the predominating configuration, but in such case the maximization of the three expressions measuring the equilibrium, and the implied minimization of *PV/T*, would occur in the adjacent system or parcels with which the considered system, or parcel, is freely connected. What is theoretically found to apply to a generic system also applies to every other generic system subject to the same conditions, in a balancing process that is entirely based on the Boltzmann factor. In other words, this means that deviations from the most probable energetic state are minimized. Finally, the third expression:

$$\Omega(E^*, V, N^*)e^{-\beta E^* + \beta \mu N^*},$$

shows that, at equilibrium, which is characterized by the most probable internal energy (E^*) and by the most probable number of particles (N^*), the entropy of the system (weighted by $e^{-\beta E^* + \beta \mu N^*}$, as the number of microstates accessible to the system is balanced, through the Boltzmann factor, with the number of microstates accessible to the surroundings) closely approximates the value of the entropy of the corresponding microcanonical ensemble, which is the maximum value possible at E^* and N^*. That is, the weighted entropy is maximized because all the actually accessed microstates are equiprobable (this was demonstrated when discussing the Gibbs entropy). It is worthy to note that when the system is not isolated, the number of microstates has to be weighted by the Boltzmann factor to account for the balancing process between the system and its surroundings. The effect of the size of the system is thereby cancelled out and only the overall effect of the system's equilibration is thus considered.

Thus, the first expression identifies the most probable macrostate, *i.e.* the equilibrium, at which no net force is available for doing work inside the system; the second shows the system has a strong probabilistic tendency to stay in such a state (the minimization of the probabilities of the microstates with energy higher than the basal level involves the maximization of the probabilities of the microstates with energy equal, or very close, to the basal level); whereas the third term demonstrates that the frequencies of the microstates that belong to the predominating configuration are all essentially equal as a consequence of the system's persistence in a macroscopic equilibrium state. Therefore, the multidimensional volume in phase space (aka hyper-volume), corresponding to the theoretically accessible microstates of the dominating configuration of the system, is entirely accessed (as demonstrated by the second term), and the system's trajectory in phase space in the long run must visit with equal frequency all the points of the region representing the dominating configuration of the system (as implied by the third term). This is just the assumption of equal *a priori* probabilities introduced to deal with the theoretical aspects of equilibrium conditions: weighting microstates for their energetic probabilities equates to give their frequencies over time. Thus, the equation:

$$\sum_m e^{-\beta E_m + \beta \mu N_m} \approx \Omega(E^*, V, N^*) e^{-\beta E^* + \beta \mu N^*}$$

means that the microstates that are accessible given the state functions of the system (*i.e.*, by hypothesis) and the microstates that are probabilistically accessed, according to the second law, are numerically equivalent when the proper Boltzmann factor is considered. Once the number of accessible microstates is weighted, there is then no practical necessity to distinguish between accessible microstates in the sense that they are possible or in the sense that they are highly probable.

Noticeably, the temperature and the chemical potential are kept fixed (in the canonical and grand canonical ensembles, respectively) by coupling the system with a large bath (reservoir or surroundings) at a given T. Since this allows the existence of real fluctuations [9], it does not exclude that these fluctuations can be macroscopic, it just says that it is extremely improbable that they be so. This means, however, that the thermal energy of a microstate of the system might indeed deviate widely from the energy that corresponds to the temperature of the bath, and then for an instant the temperature of the system would indeed fluctuate away from the fixed temperature. Anyway, the temperature is an aggregate statistical property of the system ensemble, as only the ensemble average velocity of all molecules, or the average velocity of a single molecule over long periods of time, provides a proper estimation of the temperature [9]. Actually, at the microscopic level there is not a property that is homogeneous among the particles, even though the system is equilibrated at a fixed temperature. On the contrary, at every temperature far from absolute zero, the velocities of the particles widely vary according to the Maxwell-Boltzmann distribution and it is only at a macroscopic level that the agglomerate distribution of velocities appears as a uniform property of the system. Clearly, this means that, at an almost microscopic level, inhomogeneities (between parcels that include small numbers of particles) can be very frequent. In this sense, when we say that a body has a given temperature, we are saying that such a temperature is closely representative of the temperature of that body throughout its macroscopic parts and for an appreciable time, not that it is the absolutely exact temperature of the body without any spatial and/or temporal fluctuation. This is indeed an important caveat for the thermodynamic view of temperature. The same holds true for the concentration and pressure[50] of a system that is in equilibrium with a large bath at fixed T and μ or at fixed T and P; even at equilibrium they can fluctuate through space and time, but almost always only by microscopic amounts. What is really kept fixed, at least under ideal conditions, are the number of particles and the internal energy of an isolated system, or the volume of an isochoric one. For almost all the other state functions, we are speaking of values that are representative of the system with very close approximation. It is thanks to the Boltzmann factor, and then to the second law of thermodynamics, that we can be so confident that the values we are speaking of are real properties of the bodies we study.

It is worth to note that, since the relative fluctuations for either E, V, or N decrease as the square root of the size of the system [1], in large systems that have fixed V, T, and μ by being in equilibrium with a very large bath that maintains these state functions (*i.e.*, in a grand canonical ensemble), if T, and only T, should be reset in the bath, then the parameter PV/T, which is the exact function that is minimized throughout the system, would be closely approximated by the Helmholtz free energy calculated for the same system assuming canonical conditions, and the latter, therefore, could be looked at as another function that is minimized at equilibrium. In fact, fluctuations in the number of particles would be minimal, just like those in energy once the thermal equilibrium is reached, because chemical potential and thermal energy can equilibrate independently.

Analogously, the Gibbs free energy could be used as a valid approximation in evaluating the approach to equilibrium of an open system if both T and P, but not μ, were reset in the bath. In fact, thermal and mechanical interactions, but not diffusive ones, would occur. Even though particle could diffuse, they actually would not, at least for what can be observed at a macroscopic level, because the chemical potential would remain in equilibrium. In general, if a state function is kept fixed in a system by the equilibrium with a bath, when another state function is changed (in the bath) and is completely independent of the former, then the equilibration of the bath with the system occurs according to the maximization of the overall entropy (as always), which is measured by the minimization of parameters[51] that can be either an exact measure, given by the proper parameter for the ensemble, or an approximate measure that can be borrowed from another ensemble wherein the presently equilibrated state function is theoretically, rather than just actually, fixed. For large systems, this approximation is very close, because discrete changes in one state function are much more important than equilibrium fluctuations in another one, and the latter can therefore be neglected.

CHAPTER 11

Entropy Increase as Tendency: Drive and Effector

Abstract: A useful definition of entropy is "a function of the system equilibration, stability, and inertness", and the tendency to an overall increase of entropy, which is set forth by the second law of thermodynamics, should be meant as "the tendency to the most probable state", that is, to a state having the highest equilibration, stability, and inertness that the system can reach. The tendency to entropy increase is driven by the probabilistic distributions of matter and energy and it is actualized by particle motion.

Keywords: Entropy, Equilibration, Stability, Inertness, Most probable state, Spreading and sharing, Tendency.

Although it is clear that the entropy increase is the tendency of both particles and energy to spread spatially from less equitable to more equitable distributions [33], it also needs to be clear that such a tendency is not a teleological function, that is, spreading does not occur with the aim of suppressing non equitable distributions, even though that is a result. Deviations from the most probable distributions tend to be suppressed only because they are less probable, so that the system attains the most probable distributions of particles and energy, since they are most probable, and then tends to mostly fluctuate over the dominating energy configuration.

Thus, the most probable state of a system is determined by the uniform spatial distribution of particles (eventually modified by force fields or other constraints) and the Maxwell-Boltzmann distribution of thermal energies among particles, i.e. the most probable distributions of matter and energy, which are fully combined in the multidimensional phase space. These distributions are the drivers for the function of spreading and sharing, that is, of entropy. This is why, in addition to suggesting that a useful definition of entropy is "a function of the system equilibration, stability and inertness", I also suggest that the tendency towards an overall increase of entropy set forth by the second law of thermodynamics should be meant as "the tendency to the most probable state", that is, to a macroscopic state whose distribution of matter and energy is maximally probable (according to the probabilistic distributions of matter and energy, and also considering the eventual presence of constraints), thus, with time, a system settles into the most equilibrated, stable, and inert condition that is actually accessible. Indeed, Ben-Naim [34] observed that a probabilistic interpretation reduces the second law to a matter of mere commonsense by saying that any system will go from a relatively low probability state to a high probability state.

By these definitions, the fact that in the physical world there is a tendency towards a state that is most probable is at the basis of the concepts of equilibration, stability, and inertness. In turn, it is the existence of the latter (equilibration, stability, and inertness) that allows us to think about a function that measures them. In other words, the second law of thermodynamics is the fundamental, empirical fact from which the ideas of equilibrium and entropy arise. Even the definition of Ω, at the basis of the Boltzmann entropy, implies the concept of accessible microstates; but theoretical accessibility, in turn, implies that the accessible microstates are only those constrained to the dominating configuration, that is to an arrangement of the particles and particles' energies that conforms to the probabilistic distributions of matter and energy across every macroscopic parcel of a system, which corresponds to the condition of equilibration, stability and inertness. So, the statistical definition of entropy actually implies the maximization of entropy, that is, the second law. Indeed, Gibbs [35] noticed the not immediately intuitive nature of "the notion of entropy, the very existence of which depends on the second law of thermodynamics". To state the second law in terms of an universal increase of entropy seems then a circular definition (though it is a self-consistent statement, once its meaning is grasped), whereas stating it as "the tendency to the most probable state" overcomes this trouble.

Indeed, the Gibbs equation underscores that the entropy would be maximal if all the microstates had the same probability, thus that overall equal probability is the real end state toward which there is spreading of energy, as well as of particles. Thus, I suggest specifying that entropy is a function of the tendency for particles and energy to spread to attain equiprobable macrostates over connected systems, that is, macrostates fluctuating over equiprobable combined microstates, which have the most equilibrated probabilistic distributions of both particles and energy given N, V and E or T, and as affected by any eventual constraint. Saying that entropy is a function of the tendency to attain macrostates equiprobable between combined systems is equivalent to saying that every system tends to the most probable macrostate (which is reached at equilibrium), since, as seen, each system can be considered as the combination of its parcels, or parts. Analogously, to say that a system tends to the most probable macrostate is to state that, at equilibrium, it stays at the most probable macrostate, because it fluctuates over equiprobable microstates and therefore does not have any possibility of autonomously increasing its macrostate probability further. In fact, in this condition there is not a single more probable microstate that could be accessed or where the system would be trapped by a greater stability and inertness. The system stays in the most probable macrostate because there is not a better energetic accommodation available, and then there is no force to push it out of such equilibrium. As already seen, equilibration, stability, and inertness are all different aspects of the same physical condition.

Since every large system can be considered to be composed of parcels, if all the macroscopic parcels of the system are equilibrated to the same most probable condition, then they are also equiprobable (in energetic terms, as quantified by the Boltzmann factor), and the whole system is then equilibrated to the most probable state (as a consequence, at equilibrium, connected systems have equiprobable macrostates corresponding to the most probable state of the combined system). This implies the absence of differential effects between the intensive properties of the parcels of the system[52], which occurs only at equilibrium as affected by eventual constraints that can modify the most probable distributions in different parts of the system. Hence, every system that originates from the combination of previously separated systems tends to assume a most probable configuration that minimizes the energy of each of its parts and is uniform throughout all the parts (in terms of probability, and according to the eventual presence of constraints). As the most probable state is the equilibrium one, and therefore all microstates are practically equiprobable, the tendency towards the most probable macrostate translates, at the microscopic level, to the tendency for equiprobable microstates. The energetic levelling conveyed by the Boltzmann factor assures that these two tendencies are essentially equivalent (Fig. **3**) and therefore they are equivalent aspects that characterize any entropy change at different levels.

As already noted [36], an increase in the thermodynamic entropy during a process is enabled by the motional energy of molecules, though entropy increase is only actualized if the process makes available a larger number of arrangements for the system's energy, a final state that involves the most probable distribution for that energy under the new, final equilibrium conditions. Thus, the two requisites, namely momentum energy and probability, are both necessary for thermodynamic entropy changes, but alone neither is sufficient [36]. Indeed, the tendency to increase entropy is driven by the probabilistic distributions of matter and energy, as characterizing our universe, but this tendency actualizes only if there is particle motion, which then is the effector of any increase in entropy.

How the tendency for entropy increase is driven by the probabilistic distributions of matter and energy, and how this tendency is realized by particle motion, which is therefore the effector of any increase of entropy, is discussed in more depth in the next section.

CHAPTER 12

The Probabilistic Driver and the Role of Restraints

Abstract: The probabilistic distributions of matter and energy give the direction toward which processes spontaneously tend to go. However, every existing process is presently in force because of some kinetic barrier or restraint that operated in the past: were it not for kinetic restraints (and constraints that were eventually removed by past processes), all the ongoing processes, and their causal antecedents, had already occurred billions of years ago. The role of constraints and restraints is therefore fundamental to establish the actual direction a real process can take, as well as its velocity.

Keywords: Absolute minimum, Activation free energy, Earth's thermal gradient, Free energy, Kinetic barrier, Local minimum, Metastable equilibrium, Restraints, Stable thermodynamic equilibrium, Thermodynamic equilibrium, Transition state.

Through every spontaneous process, a system ultimately reaches a more stable thermodynamic equilibrium, with an unchanging, maximally probable macrostate and fluctuating microstates, and it will spend a larger fraction of time in those microstates that have a larger probability [34, 37]. As seen, a real system at temperature T fluctuates over the microstates of the dominating configuration, the ones having larger probability, for approximately equal times and eventually over microstates outside of the dominating configuration but with probability that is then approximately equal to zero [9]. Indeed, a thermodynamic equilibrium exists and is reached just because it is the maximally probable macrostate. It is, in fact, implicit in the concept of probability that if a system can be in a microstate that is not very probable, that just means that the system will spend less time in that microstate, and if the system and its surroundings had macrostates with diverse

Alberto Gianinetti

overall probabilities, and were not isolated from each other, particles and/or energy would make them equiprobable by spreading[53].

It should be noted, however, that a system may stay for a very long time in a state of metastable equilibrium corresponding to a local minimum of G (the form of energy that is minimized at the stable equilibrium of a closed system at constant T and P and which maximizes the overall entropy), or of a different energy function appropriate for different conditions [9]. This may happen because the energy barrier separating this local minimum from the absolute minimum (corresponding to the most stable equilibrium) can be large enough to be almost insurmountable by random fluctuations in the system [38]. A system that resides at a local, but not global, free energy minimum is said "metastable". The existence of a free energy barrier (actually, a transient state with higher free energy) between a metastable and the globally stable state of the system enables the former to exist for periods of time that permit it to be investigated. Ultimately, however, metastable systems will attain global equilibrium if given infinite time [9]. In the case of rates of chemical reactions, commonly, there is a barrier in free energy (F, if the ensemble is not specified) the system must overcome. This barrier, addressed as $\Delta F^{\ddagger}$, determines how fast the reaction can proceed [9]. The barrier is often called the activation free energy. The state of the system when it is at the height of this barrier is called the transition state (which is the rate-limiting state for a reaction, since this state has the highest energy and therefore the lower probability of occurrence), and it is usually denoted with a superscript symbol "$\ddagger$". After reacting, if the process is energetically favourable, there will be a net lowering in free energy at the product state [9].

It is even possible for a system to remain in a kinetically-arrested metastable configuration for very long times if the minimization of energy involves extremely slow particle motions (*e.g.*, the deformation of ordinary glasses under the influence of gravity takes centuries) [9, 38]. Every existing process is presently in force because of some kinetic barrier, or restraint, that operated in the past; were it not for kinetic restraints (and constraints that were eventually removed by past processes), all the ongoing processes, and their causal antecedents, had already occurred billions of years ago. Purposeful removal of kinetic barriers (*i.e.* restraints) and constraints is the means by which we can intervene with spontaneous processes (*i.e.*, processes granted by an entropy increase) for shepherding, steering, and channelling them into a desired direction. Indeed, constraints, like the walls delimiting a gaseous system (Figs. **1** and **2**), allow recovery of some energy from spontaneous processes, that is, from entropy increments. By itself, entropy is not a force; it just measures the probability of a state, and therefore the equilibrium of a system, that is, the unbalancing of forces. It is from the unbalancing of forces that we can obtain work.

In the long run, kinetic restraints are responsible of the fact that stars still shine; they will continue to glow until the nuclear reactions that keep them hot are exhausted. A kinetic restraint is also responsible for the persistence of a thermal gradient in the Earth's interior. Although some heat flows from the Earth's interior to the surface and is then irradiated to the atmosphere, the Earth's crust effectively acts as an insulating barrier, thus that quasi-steady-state conditions can be considered to occur for the Earth's thermal gradient [28]. Indeed, the vast majority of the heat generated in the Earth's interior is attributable to the decay of radioactive isotopes within the mantle, and the amount of heat produced by this radiation is almost the same as the total heat emanating from the Earth, which means that the Earth is cooling very slowly. Interestingly, although this radioactivity-linked heat is generated in the mantle, the core is hotter, *i.e.* there is a stationary thermal gradient between the hot mantle and the hotter core. This thermal inhomogeneity of the Earth's interior has been suggested to be linked to the pressure-induced freezing of the metallic fluid of the outer core (which is fluid because the high temperature dominates high pressure) on the inner solid core (which is solid because the very high pressure raises the fusion point above the actual temperature). This is a mechanism that provides latent heat that promptly diffuses through the metallic core and maintains hotter isothermal conditions in the Earth's deep interior [28]. The enlargement of the solid core is due to the fact that, although very slowly, with time the Earth cools and the surface of intersection between the geothermal gradient and the curve of the pressure-induced melting progresses farther from the Earth's centre, causing the inner solid core to grow outwards. The heat thusly generated diffuses much more rapidly throughout the metallic core, which has a higher thermal conductivity, compared to the mantle, which has a low thermal conductivity (*i.e.*, it acts as an insulator). So, apparently because of the different kinetics of thermal transfer in the core and in the mantel, a quasi-static thermal energy in the Earth's interior is associated with a persistent thermal gradient. Until there is a fluid outer core that solidifies, the Earth's core will be hotter. Some tiny amount of heat might also be released as a consequence of gravitational sedimentation, a gravitational sorting process by which the denser, heavier components of the fluid mantel are drawn to the gravitational centre and then accumulate at the core, while the less dense materials are displaced outwards [28]. Heat is then produced because of the friction created by this process (which mostly occurred when the entire Earth was solidifying) as matter redistributes to a lower energy configuration, specifically with denser materials on the bottom. Some heat may thus still be generated in minimal amounts since some iron and nickel are still present in the crust, and then they represent a small source of gravitational potential energy. Of course, both these supposed mechanisms, as well as each physical process happening in the universe, are driven by an entropy increase. The latter is therefore associated with every process, whether it causes uniformity or diversification, produces order or disorder, or generates or destroys complexity. In fact, whereas entropy is mathematically defined, these other concepts are not so sharply measurable and comparable. If order and complexity exist, and they were not already present at the origin of the universe, it is because some processes generated them. Those processes, whatever they may be, were associated with an increase of entropy.

Coming back to consider the conditions that define the equilibrium, which will eventually be achieved in every system after long enough time (at least by hypothesis), the positional partition function and the Maxwell-Boltzmann theoretical distribution establish which actual distributions of the particles are most probable throughout both space and energy, and therefore they define the equiprobable macrostates toward which every system has a tendency, as measured by entropy. By neglecting interactions, entropy can then be seen as having two components: a positional one and a thermal one[54]. Hence, an isolated system under equilibrium conditions is expected to observe a spreading of particles throughout space and energy levels (or thermal states) according to the positional and thermal partition functions, which represent the probabilistic distributions of the microstates. This is the condition of maximum entropy attainable by the system and corresponds to the Boltzmann entropy[55] $S(E,V,N) = k_\mathrm{B} \ln \Omega$ [8]. However, a system that is not at thermodynamic equilibrium, for example because it is closed and receives (and/or loses) energy from (to) the outside, presents at least some microstates that are favoured and therefore come to be more probable than these distributions predict. This causes deviations from the probabilistic equilibrium that maximizes the entropy of the system. These deviations are assessed by the Gibbs equation that, consequently, calculates a reduced value of entropy for the system. If, for example, we have a system that is at dynamic equilibrium, *i.e.* it simultaneously receives and loses roughly the same amount of energy from/to the outside (like the Earth) thus its average temperature (thermal energy) can be considered to be constant, and neither the number of particles (as the system is closed) nor the system volume change[56], then the net (negative) departure of the system entropy from the equiprobability assumption is measured as:

$$-\Delta S = -k_{\mathrm{B}}\Sigma_{m}\, \wp_{m}\ln\, \wp_{m} - k_{\mathrm{B}}\ln\Omega$$

that is, by the difference between the Gibbs and the Boltzmann[57] expressions. Consequently, the intensity of the tendency towards equiprobable macrostates (higher entropy) is:

$$\Delta S = k_{\mathrm{B}}\cdot(\ln\Omega + \Sigma_{m}\, \wp_{m}\ln\, \wp_{m})$$

This difference, or, more generally, the possible increase in Gibbs entropy between two macrostates of a system, is the probabilistic drive that defines the intensity of the tendency towards equiprobable macrostates.

CHAPTER 13

The Motional Effector

Abstract: According to the second law of thermodynamics every spontaneous change, or process, is associated with an increase in entropy. Although the probabilistic distributions of particles and energy give the possible direction of a process, its occurrence is enabled by the motional energy of the particles. Even particles, however, are subjected to constraints of motion that slow down the attainment of some possible changes and thereby reduce their probability of occurrence, especially if alternative pathways to increase entropy are possible and can be accessed faster. Kinetic restraints are therefore key determinants of which processes are activated among the different possible ones.

Keywords: Average kinetic energy, Constraints of motion, Energy redistribution, Energy spreading, Particles collisions, Particle motion, Zeroth law.

The Maxwell-Boltzmann distribution, by defining the probability distribution of the particles among thermal states (or nondegenerate energy levels), also determines the average level of thermal energy of the particles, which is expressed as $k_B T$. In fact, a single parameter, $\beta = 1/(k_B T)$, determines the most probable distribution of the particle population over the thermal states of the system, and thus the thermodynamic temperature is the sole state function governing the most probable distribution of particles among energy levels of a system at thermal equilibrium [24]. Temperature (multiplied by k_B) represents the overall level of thermal energy that is redistributed throughout all the particle collisions (assuming that collisions are elastic, *i.e.* at each collision the total thermal energy of the colliding molecules is conserved) and is therefore a variable defining the state of the system. In the collisions, there is a tendency for energy to be transferred from

Alberto Gianinetti

the faster particles to the slower ones, so the energies even out [1]. In fact, collisions result in the ceaseless redistribution of energy not only between the molecules, but also between the quantum states that each molecule occupies, including their different modes of motion [29]. Temperature determines the motion of the particles and, therefore, it also determines the overall frequency of the collisions in the system. In fact, for a given type of particle, the thermal energy of these particles is exactly proportional to their translational energy (that is, the kinetic energy of the particles), and this energy is in turn proportional to their velocity, which thus is proportional to the thermal energy[58]. At a given (by the combination of N and V) overall density of particles, their average velocity determines the overall frequency of the collisions in the system. Indeed, in the configurational partition function the effect of temperature is included, but, differently from the thermal partition function, in this case the effect is not linked to the distribution of particles through thermal states (or nondegenerate energy levels), rather it has the peculiar role of affecting the probabilities of a system being in the various microstates [1]. Hence, the temperature determines the average translational energy of the particles, which diffuse throughout the space because they have translational energy. In addition, thermal energy causes motion, and particularly translation (at least in the case of fluids at low particle densities), which is the main reason why particles collide. In each collision, the colliding particles share their thermal energies, which are immediately redistributed in a probabilistic manner, thus assuring the transfer of thermal energy (energy spreading) as a function of the motion associated to that energy. Energy spreading is therefore strictly linked to energy sharing.

The collisions determine the sharing of thermal energies from the colliding particles; energies that are afterwards probabilistically divided among the particles that collided, according to the Maxwell-Boltzmann distribution, which is roughly mean-centred. That is, whatever the energies of two colliding particles were, their aggregated energy is redistributed according to a function that gives a maximum probability approximating an equitable repartition. In this way, the populations of states remain almost constant, but the precise identities of the molecules in each state may change at every collision [25]. This is the core mechanism of the tendency for the sharing/spreading of energy. Actually, the Maxwell-Boltzmann distribution is skewed toward higher energies, and therefore the mean speed of the particles is slightly higher than the most probable speed (and hence the most probable translational kinetic energy), but, anyway, the fractions of particles with very high and very low speeds will be very low [25].

Thermal energy, and particularly translational energy, therefore appears to be the cause of particles spreading throughout space and of energy spreading among particles (*i.e.*, sharing). Hence, thermal energy ($k_B T$), and particularly translational energy (at least in fluids, in solids, vibrational motions take a prominent role) is the effector (or enabler) of the tendency toward equiprobable macrostates. In fact, phase transitions from more constrained to less constrained states are accompanied by massive and discontinuous increases in the entropy because during the transition the molecules of solids, liquids, and gases rapidly gain greater freedom to move around, facilitating the spreading and sharing of thermal energy. Notably, in addition to atoms and molecules, spreading and sharing of energy can also be mediated by irradiating photons (electromagnetic waves) and other massless particles. In any case, it is the movement of particles that essentially causes energy spreading, and it is the interaction of particles, collisions, or emission/absorption, that determines energy sharing.

As a consequence to what was discussed above, the increase of entropy at higher temperature is not directly due to the higher average kinetic energy (a tendency to higher energy levels would be a tendency towards an increase in temperature, which is not the case), but it results from the fact that at higher temperatures: (i) the Maxwell-Boltzmann distribution of thermal energy is more spread (toward quantum states of both lower and higher energies); and (ii) the de Broglie thermal wavelength of the particles (Λ), which represents, for an ideal gas, the space occupied by a atom (intended, in quantum mechanics, as the spatial extent of the particle wave packet), decreases with increasing temperature. In fact, with increasing temperatures the wave-packet volume, Λ^3, decreases and the particles have a bit more space over which to spread and arrange in the system volume, and thus positional entropy is higher [19, 20]. So, an energy gain at fixed volume causes gas atoms to increase their possible arrangements (microstates) in the system [19]. This provides an alternative explanation for the work that is lost when the system settles into a new equilibrium state following an increase in temperature. In addition, higher thermal energy makes particles to move faster and collide more frequently, and this translates into faster spreading of both particles and energy through space, that is, higher temperatures also enforces the actualization of entropy, which is enabled by the kinetic energy of the particles. Indeed, the spreading of energy requires that there is energy to be spread and thermal energy is the kind of energy that can quickly spread from, or through, most systems. Therefore, at absolute zero, entropy is also zero because there is no thermal energy to spread. Above this temperature, the tendency for energy spreading is augmented by increasing thermal energy. The higher entropy occurring in an isolated system at equilibrium by the effect of higher temperature can be seen as an increase in the tendency for the particles and energy to spread through space in consequence of the higher thermal energy. From this point of view, the tendency towards an overall increase in entropy can be seen as a thermal-dependent, probabilistic tendency to spread, in terms of higher particles speed and higher frequency of particles collision, and the consequent faster spreading of energy through space.

On the other hand, particles are subjected to constraints of motion that reduce the number of available quantum levels and then the number of accessible microstate, that is, entropy. These constraints may correspond to [22]: (a) constraints on the number of independently moving particles; *i.e.*, on whether the particles must move as an aggregate or can move separately; (b) constraints on the direction of motion; (c) constraints on the volume in which the motion is executed. Molecular bonds and a solid state are constraints of the first kind. For the other two types of motion constraints, physical barriers are obvious ones, but, unless they are insulating, energy can spread through them. Gravity and other fundamental forces are constraints as well. Actually, gravity represents a restraint on the motion of particles in a centrifugal direction with respect to the gravitational field. Molecular bonds are constraints that reduce the number of particles in the system and therefore they reduce the number of microstates that are accessible to the system, *i.e.* its degrees of freedom[59]. In addition, whereas atoms do not have a rotational component for thermal energy[60], molecules do have one (as they do not have a perfect radial symmetry), and this fact reduces the two other components of thermal energy, specifically translational energy, which is much more effective in causing the tendency towards equiprobable macrostates (or to the most probable overall state).

In a more general view, as already noted, entropy could be defined "a function of the system equilibration, stability, and inertness", and the tendency towards an overall increase of entropy set forth by the second law of thermodynamics should be meant as the probabilistic tendency towards energetic uniformity (that is, tendency to equiprobable macrostates) of adjacent, but not isolated, systems in the absence of constraints. It is worthy to note that this definition of entropy recalls both the second law of thermodynamics (the entropy of an isolated system tends to increase, and this is an obvious consequence of considering an increase of entropy as a physical, probabilistic tendency) and even the so-called zeroth law (if two thermodynamic systems are each in thermal equilibrium with a third, then all three are in thermal equilibrium with each other), since thermal energy tends to be uniform over systems. This occurs by means of particles collisions in isolated and closed systems, and also because of particles translation in open systems. Indeed, any flow of heat is not a tendency to equalize energy, but to equalize entropy. In fact, temperatures are equalized, not energies [8]. Energy shows a net flow only if it leads to an increase in the overall entropy. Actually, persistent macroscopic departures from uniformity in $k_B T$ over a system can be seen as non-null interactions between positional and thermal partition functions. In fact, such departures would mean that there is some correlation, along one or more spatial dimensions, between the positional partition and the Maxwell-Boltzmann distribution[61]. A correlation between thermal and positional partitions would mean that some parcels would have higher levels of energy, whereas any high-energy spot would be hugely improbable just because of high energy (Fig. **3**). Indeed, particles collisions assure that, at thermodynamic equilibrium, this does not occur, as sharing of energy at each collision and its subsequent probabilistic redistribution causes the levelling of thermal energy throughout the system.

Things in the universe change because particles (including irradiating photons and other massless particles) move, and probability drives the movement in a preferred direction. As a corollary, perfectly reversible processes should randomly oscillate in one or the other direction in a dynamic equilibrium since there is no preferred direction.

CHAPTER 14

Spreading & Sharing is a Common Outcome of a Physical Function Levelling Down Available Energy

Abstract: Entropy has been defined as a probabilistic function of energy spreading and sharing, and most often this description provides a straightforward way to conceptualize entropy. It is shown that, more in general, the spreading and sharing of energy is a common outcome of a physical function levelling down available energy. The latter, as formulated by a mathematical term called the "Boltzmann factor", originates from the equilibration of forces at the microscopic level and is effected by the net levelling force that results as statistical outcome of all the microscopic forces and that always pushes the system towards the dynamic equilibrium. This net levelling force is the rationale for which work can be done at a macroscopic level, and its derivation from the microscopic world explains why it is linked to equilibration and therefore to entropy increase.

Keywords: Energy dissipation, Free energy, Gradient equalization, Levelling function, Living organisms, Macroscopic homogeneity, Microstates equiprobability, Minimum possible level, Most probable distribution, Spreading and sharing, Unused heat, Work as change.

So far, entropy appears as a probabilistic phenomenon of sharing and spreading of energy, as underscored by Leff [19, 20], and we can therefore continue to use such a metaphor for now. Indeed, in each microstate of the dominating configuration the position and momentum of each particle are equilibrated according to the most probable distribution of matter and energy throughout the whole set of particles of the system. As seen, the fact that microstates with energy levels deviating from that of the dominating configuration are improbable is equivalent to the fact that the macrostate shows a distribution of matter and energy that, at a macroscopic level, is uniformly probable throughout the whole system. Any nascent inhomogeneity would be immediately prevented from developing further by the fact that either matter or energy, or both, would tend to flow or spread from/to that part to/from the others. Thus, the fact that the system fluctuates over microstates with probabilities strongly opposed to their energy levels (Fig. **3**) translates into a macroscopic homogeneity for the system in terms of probabilistic distribution of matter and energy, *i.e.* into equiprobability.

The fact that real processes are continuously observed demonstrates that equilibration is ongoing everywhere. Notably, it is exactly the existence of non-equilibrated conditions, or potentially improbable distributions of matter and energy, that allows for the obtaining of work (as stressed in classical mechanics) when (by dissipating some potentially available energy[62]) a constraint is removed. In other words, the tendency towards the most probable state (*i.e.* to equiprobable macrostates, if we consider the process of combining two systems) can be exploited to produce work when the degrees of freedom of a system are increased. If the removal of the constraint is casual, the produced work is aimless[63]; if, on the other hand, the removal of the constraint is purposeful, the obtained work can be functional to some aim. Indeed, an entropy increase is the material cause of any change, but whereas in many natural processes it is also the only cause, in purposeful processes, purposefulness implies that the process could be started or not, at a given time, by choosing the appropriate constraints upon which to act (that is, purposeful choice can be one of the causes of an event, if we assume that free will exists[64]). It is just by coupling opportune processes that can produce some work with other processes that can use at least part of that work, that purposeful changes can happen. Knowledge and choice are then key features of purposefulness, but an increase in thermodynamic entropy is always a fundamental requirement in the realization of every material purpose. Even single processes that, by themselves, are spontaneous can be reversed by coupling them with another process that increases the overall entropy more than the former reduces it. Everyday some fuel must be consumed: to remove heat that has leaked inside a refrigerator, or to put back on the shelf a fallen book, or to put air back into a tire once it has leaked out [1]. More generally, complex systems can exploit existing disequilibria to obtain work for carrying out some function, purposely or not.

It is indeed by this mechanism that, in the current era, living organisms proliferate on Earth[65]: life can be considered as a metastable condition associated with a reduction of the entropy on Earth, since living organisms maintain and multiply as constraints to the free spreading and sharing of their own molecules and energy (*i.e.*, they reduce the degrees of freedom of their matter and available energy). In fact, it has been argued that the source of all the complexity of life might just be nature's tendency to equalize things, and, accordingly, every living system is made to catch, store and degrade gradients [39]. Anyway, life is maintained essentially because of plants' capability to transform some high-energy radiation (sunlight) into low-energy radiation (Earth's infrared emission). Specifically, photosynthetic organisms intercept some sunlight and transform part of it into heat irradiating into the space, and obtain, in this irreversible process[66], some chemical work stored as potential energy in the usable bonds of organic compounds. So, until the sun will emit light, *i.e.* until it will guarantee non-equilibrated conditions, plants will have the opportunity to destabilize the local minimum of energy that constrains solar photons into their high-energy state, thereby producing a larger number of low-energy infrared photons [40]. By thus increasing the degrees of freedom for solar radiation, plants exploit the tendency to the most probable, low-energy state and manage to produce chemical work to maintain life. Increasing the degrees of freedom of solar radiation allows them to decrease the degrees of freedom for some of Earth's matter, although to a lesser extent.

It can be recalled that in classical mechanics the interest is focused on the production of work. Cyclical processes are typically used to this aim. These processes exploit macroscopic discontinuities of matter (*e.g.*, ΔP) or energy, for example ΔT and Δh, the latter indicating some difference in the potential energy of an amount of matter subjected to a gravitational field, or other equivalent discontinuities or disequilibria. The maximum yield of work that can be obtained by these processes is proportional to the intensity of the discontinuity (and can be reached only under reversible conditions, assuming the discontinuity is unaffected during the process). An example is a heat engine where the maximum yield of work is $\eta = 1- T_c/T_h$, wherein T_c is the temperature of the cold reservoir and T_h is the temperature of the hot reservoir. This expression shows the maximum theoretical thermal efficiency of a heat engine is highly dependent on the coldness of the cold reservoir, and it can be 100% only if such a reservoir is at the absolute zero. As noted by Lowe [41], the absolute zero is the temperature to which the kinetic energy must fall in order for all of it to be converted into work. This is because, in general, a given high level of potential energy can be exploited only if there is also a lower level that is accessible for a transfer, and the given high level can produce as much work as the lower level is closer to its minimum possible level. Consider, for example, that the potential energy obtainable by a mass placed at height h = x-x_0 in a gravitational field increases, keeping x (the position of the mass) fixed, for decreasing values of x_0 and is maximal when $x_0 = 0$ (for a gravitational field, the absolute zero would be the centre of the field, which is, however, usually inaccessible). In the case of temperature, the minimum possible level is the absolute thermal zero (0 K), which is also difficult to access; whereas, for a pressure gradient, the minimum is an empty space, which is at least easier to approximate. In any case, it is clear that the work yielded by the discontinuity is maximized (theoretically 100%) only if the reference low-energy level is at its absolute zero. As noticed by Stoner [4], this is the reason for the limited availability of potential work in thermal engines operating under common conditions.

For each state function, the most probable, stable, and inert condition is the one wherein any well of low entropy is filled up, that is, any discontinuity or disequilibrium is levelled off. This is why the maximal yield of work is obtained by a reversible process that exploits a low-energy reservoir that is as close as possible to its theoretical minimum. Nonetheless, a high yield can be obtained through a reversible process that exploits a very wide discontinuity, *i.e.*, a high-energy system coupled with a much lower-energy reservoir. For processes that cannot exploit a low-energy reservoir that is at its theoretical minimum (*i.e.* most real processes), the yield of work is always less than the maximal work that could be obtained from a high-energy reservoir, even if a reversible process can use the available discontinuity with perfect efficiency. In a reversible process based on heat transfer, the internal energy that does not produce work (owing to a low-energy reservoir that is above its theoretical minimum) is therefore merely unused heat, not waste heat. In fact, no potentially available reversible work has been lost. It is simply unusable under the given conditions.

In general, some internal energy of a macroscopic system can be dissipated, that is, transformed into a less utilizable form (*i.e.*, it is no longer available for doing work), but the internal energy of a system can never be diminished unless if it is transferred outside the system. So, the inherent spontaneous tendency is not just to a lower level of internal energy, but to a lower proportion of utilizable energy, *i.e.* to a more spread and shared distribution of energy and matter that makes the energy less utilizable and the universe more stable and probabilistically equilibrated. Typically, an increase of entropy means a probabilistic levelling of energy and work can only be "extracted" from the internal energy when it is available as potential energy, that is, in the form of a gradient, or discontinuity, of matter or energy. This provides the connection between statistical and classical thermodynamics (which can be exactly formulated in terms of the Boltzmann factor, as previously seen). In this sense, entropy can be meant as a measure of equilibration for the forces acting in the system; forces that are driven by the probabilistic distributions of particles and energy, and by their interaction with existing constraints, and that are enabled by the motions of the particles. Free energy is then promptly interpreted as a measure of the work that can be produced by the net force that results from any imbalance in the system.

The physical function levelling down available energy can provide work because the levelling of forces at the microscopic level is effected by the net levelling force that obtains as statistical outcome of all the microscopic forces. The dynamic equilibrium is thereby provided. Spreading and sharing of particles and energy is therefore a common effect of the probabilistic levelling function, though not a necessary one, as we'll see. Ultimately, the physical function levelling down available energy often acts as a spreading and sharing function, and this is why the latter often can provide work.

CHAPTER 15

Changes of Entropy: The Fundamental Equation and the Chemical Potential

Abstract: Any change in the physical world is consequent to a net force levelling down available energy. Different features of a system, or process, can determine on which properties of the system the levelling force eventually operates. Examples are the equilibration of the pressure of a gas, of the temperature of bodies, of the concentration of solutions, of the position of a body with respect to a fluid in the presence of a gravitational field, and so on. All these phenomena occur because of the levelling force and are associated with an entropy increase. The entropy increase can be calculated through specific relationships.

Keywords: Accessible microstates, Chemical potential, Colligative properties, Conjugate pairs, Equal multiplicity, Extensive variables, Fundamental equation, Intensive state functions, Lost work, Multiset permutations, Overall entropy increase, Particles concentration, Reversible change, Saturation, Spontaneous transfer, Standard conditions.

THE FUNDAMENTAL EQUATION FOR THERMODYNAMIC CHANGES IN THE MACROSCOPIC PROPERTIES OF OPEN SYSTEMS

Although the change of entropy in a system, following an interaction with its surroundings (a transfer of heat, of work, or of particles), might be theoretically computed by the change in the number of microstates at equilibrium (in terms of the Gibbs and Boltzmann entropies), it is much more practical to calculate it on the basis of the actual changes in those state functions that are linked to the effective transfer(s) of heat, work, or particles. Correspondingly, the spontaneity of a transfer, specifically, the overall change of entropy associated to it, can be computed in terms of the actual changes in the state functions of both the system and its surroundings. The change in the Gibbs entropy toward its maximal value at equilibrium would provide the direct calculation of the entropy increase in terms of the overall increase in accessible microstates, that is, from a microscopic point of view. However, calculating changes in the proper fundamental thermodynamic parameter (like the Helmholtz and Gibbs free energies for the canonical and isothermal-isobaric ensembles, respectively, or PV/T for the grand canonical ensemble) provides a much more comfortable approach to compute, from a macroscopic viewpoint, the entropy changes, since, as previously shown, the two approaches are mathematically linked.

As seen, any transfer of energy or matter is caused by particle motions and its net effect is directed by the probability of the actual distribution of energy and matter in the multidimensional phase space. Every spontaneous change moves toward a macrostate that is overall more equiprobable among its macroscopic parts, and therefore more equilibrated, stable, and inert. Hence, any spontaneous transfer is associated with an increase in entropy, *i.e.* a dissipation of energy, because the overall condition following the transfer shows a better probabilistic arrangement of the system(s). For example, when two isolated systems become connected, they exchange heat if their temperatures differ, exchange particles if their density is not homogeneous, and mix their particles if their compositions are not the same[67]. Thus, the two systems integrate with each other, increasing their overall stability[68] and thereby decreasing their capability of doing work, that is, some potential energy is dissipated. When this happens, the joint system has settled into a probabilistically more stable condition. This adjustment to the overall stability/inertness is what makes the system less capable of doing work, thus that the entropy increase is precisely what accounts for the dissipation of energy, as found by early studies in classical thermodynamics. Again, it must be reminded that various constraints, interactions, and force fields can affect the probabilistic outcome.

From a macroscopic point of view, the changes of the entropy (at a given temperature) of a large system occur because of changes in the macroscopic state functions of the system. For an open system under grand canonical conditions (wherein T, V, and μ are fixed, which can be considered as representative of many real systems), reversible changes occur at equilibrium according to the fundamental equation [9]:

$$dS = dE/T + PdV/T - \mu dN/T$$

which is based on the macroscopic state functions and summarizes the set of partial derivatives of the entropy changes occurring in a system due to thermal, mechanical, and diffusive processes, respectively[69] (note that the last term becomes $-\Sigma\mu_i dN_i/T$ if more chemical species are present). Likewise other expressions of entropy change, the fundamental equation is expressed in differential terms. As explained when discussing the Clausius' equation, this allows to render the relationships between state functions in a form that is as general as possible and it makes calculations and measurements more straightforward. The signs of the three terms depends on the fact that if heat is added, the temperature must rise, and entropy increases; or if the volume is increased, there are more possible spatial arrangements, and the entropy increases; or if particles are added into a given volume, the system's entropy changes in relation to the chemical potential. How, actually, the system's entropy changes in the latter case is quite complex and deserves some discussion. Nonetheless, since in many ordinary systems the chemical potential is negative[70], the minus sign in the third term of the fundamental equation assures that, in these systems, the higher the chemical potential (*i.e.*, the closer it is to zero and then the lower its absolute value is) the smaller is the increase in the entropy of the system that occurs when new particles are added. Notably, in discrete terms, if particles are added into a given volume, the chemical potential rises [1].

THE CHEMICAL POTENTIAL

It is commonly noted that chemical potential can be seen as the tendency of a substance to redistribute in space, or to react with other substances, or to change physical state, decompose, rearrange, or decay [42]; all these features are indeed derived from the fact that the chemical potential can be basically defined in terms of entropy (and then in terms of spontaneity of change). In fact, chemical potential can be defined as the partial derivative of the system entropy as a function of an infinitesimal change of N at a given T [9]:

$$(\delta S/\delta N)_{E,V} = -\mu/T$$

This theoretically might be obtained in a system with rigid walls by adding (for example, through a tiny hole) one particle having a kinetic energy equal to zero. In this way, the thermal energy of the system would not be increased and the new equilibrium temperature is used in the formula (however, if the system is large its temperature would be practically unaffected). In reality, a transfer of particles cannot occur without transferring energy as well; a particle with kinetic energy equal to zero does not move, and it is therefore quite impractical to transfer. So, even apart from considering the energy of matter ($E = mc^2$), particles cannot be transferred disjointly from energy, and temperature and chemical potential equilibrate simultaneously, even though they equilibrate independently in the presence of a thermally conductive wall (or in the absence of any wall). Hence, this definition of the chemical potential is theoretical and can never become operational. However it is very useful, because μ is much easier to measure experimentally than entropy, and, therefore, changes in S can be calculated when μ is known, as occurs for heat transfers and PV work. In addition, if the correct interpretation of the derivative, in this case, is that only the marginal (infinitesimal) change of N gives the proper value of μ, then even T can be considered to be fixed for the system at constant E. It can be envisioned that, in any case, once T is established:

$$\mu \equiv -T(\delta S/\delta N)_{E,V}$$

and since the absolute temperature T is always positive (except in a very few extraordinary conditions [21]); it turns out that the actual value of μ is negative in any condition wherein the marginal addition of a particle increases the system entropy. It can be positive only when the marginal addition of a particle brings about a decrease in the system entropy.

The third term of the fundamental equation is derived from turning the above definition over (there is not circularity, because μ is theoretically derived by S, but it is easier to measure experimentally, so that changes in S can be calculated when μ is known). It is worthy to note that, as a consequence of its dependence on entropy, the chemical potential also corresponds to the amount by which each particle can change the free energy of the system to which it is added (as seen, at constant temperature and pressure the Gibbs free energy is the proper thermodynamic potential aimed at determining equilibrium conditions). Indeed, the definition of the chemical potential as the change of the system entropy following the addition of particles at fixed E and V [9] is equivalent to the definition of the chemical potential as the change in Gibbs free energy following the addition of particles at fixed T and P [25], because the chemical potential in single component systems is simply the per-particle Gibbs free energy [9]:

$$\mu = G/N$$

This change of ensemble, however, implies that changes in pressure cannot be properly evaluated when the chemical potential is dealt with in terms of the Gibbs free energy (which involves an isothermal and isobaric ensemble). Thus, the chemical potential of a system, μ, measures the average change in entropy per entering particle that is not due to the transfer of heat or PV work (since E and V are fixed by definition for μ, and T and P are fixed by definition when using the Gibbs free energy). For processes involving the two latter types of transfer, the respective terms in the fundamental equation are used to calculate the overall entropy change. Assuming ideal conditions, *i.e.* there are no interactions among the particles, the only thing that changes in the system is the particle concentration. As noted above, since in many ordinary single component systems the chemical potential is negative, the chemical potential and the system entropy often increase or decrease together.

PARTICLE CONCENTRATION AND CHEMICAL POTENTIAL

The effect of concentration on entropy can be appreciated by considering the chemical potential of an ideal gas: the number of positional arrangements of the particles determines the number of positional microstates as particle present/absent in each spatially localized partition unit, in terms of combinatory calculus [43]. So, considering the spatial partition (V/Λ^3) of a system as a finite multiset (that is, a kind of set that allows multiple instances of its elements; in this case, particle present/absent at each spatial unit), then a multiset permutation is an ordered arrangement of these elements (particle, empty) in which each element appears exactly as often as its multiplicity (the multiplicity of an element is the number of instances of the element in a specific multiset) in the partition (thus, in the partition there are N particles and $V/\Lambda^3 - N$ empty units). Then, the number of multiset permutations of the spatial partition is given by:

$$(V/\Lambda^3)!/[N!(V/\Lambda^3 - N)!]$$

The maximum number of multiset permutations occurs when there is an equal multiplicity for every distinct element. In an ideal gaseous system this occurs when the number of particles equals the number of empty units in the spatial partition. So, when a change in the number of particles in a system is considered, this is the condition at which the number of possible positional microstates and the positional entropy are maximal. Therefore, when the number of particles is lower than this ratio (like in most gaseous systems), the addition of new particles increases positional entropy, whereas if the number of particles is higher than this ratio, the addition of new particles would decrease it (which, however, usually does not occur, since the attractive forces among the particles cause part of the gas to condense into a liquid phase well before this concentration, and the corresponding pressure, is reached, at least if temperature allows it). A similar phenomenon occurs for an ideal solution, wherein the particles of the solvent take the place of the empty units. In this case, however, it must be considered that, if the volume has to be constant, each particle of solute must replace a particle of solvent; that is, the addition of solute particles means a corresponding removal of solvent particles. In practice, since concentration is an intensive state function, the ratio of particles in a parcel of given volume can be considered.

However, the positional entropy is only one component of the system entropy; in fact, the overall number of accessible microstates in phase space is $\Omega = \Omega_m \cdot \Omega_p$. As the temperature is practically unaffected, there is no change in the available momentum entropy (*i.e.*, there are no additional momentum microstates, Ω_m), but the addition of a particle, to the N already present, does increase the overall number of accessible microstates in phase space, and therefore the corresponding entropy. In other words, the thermal energy can be spread among more particles. Indeed, each new axis in the phase space (*i.e.*, each new particle in the system) should make possible Ω_m new thermal x positional combinations, and, since the exact position of the particle counts, there should be:

$$(V/\Lambda^3 - N) \cdot \Omega_m$$

new thermal x positional permutations, which at high temperatures is large enough to overcome the loss in positional entropy. For this reason, at temperatures at which quantum effects are negligible, the chemical potential of an ideal gas is [9]:

$$\mu = -k_B T \ln [V/(\Lambda^3 \cdot N)]$$

that is, it is always negative and increases when new particles are added, finally reaching zero when the entire volume of the system is filled with particles (though, at high density, intermolecular interactions become important and the system usually condenses). Because of the link between μ and S, this means that the entropy of an ideal gas never decreases when concentration increases, but the increase in entropy tends to zero as the system volume fills up. This is not the case for a solution where particles of solute are being added, because the system volume is already filled up by the solvent molecules. Since the particles of solute take the place of the solvent ones, there is no increase in the momentum x position phase space, and then the system entropy should actually decrease when solute particles are being added above an equal multiplicity with solvent. As a spontaneous decrease of entropy, *i.e.* a spontaneous contraction of the accessible phase space, would be a violation of the second law, something else actually happens, as we are going to see.

The fact that, in a solution, the condition of equal multiplicity between the particles of the solvent and solute would represent a watershed between two slopes of increasing and decreasing entropy means that, at equilibrium, there can be two possible multiplicity ratios for each value of entropy below the watershed maximum. However, if the whole system remains in the form of a solution, one ratio would imply a reduction of overall entropy once such threshold is trespassed, which cannot be the case. The solution, therefore, demonstrates saturation. It turns out that the formation of a physical state that reverses the situation, *i.e.* wherein the added particles can leave entropy unchanged, or even can increase it again, is favoured, and it is then a thermodynamic necessity. In this way, some solute particles are no longer dispersed into the solvent, but, rather, some solvent particles are dispersed into a mass of solute particles. Thus, beyond the watershed maximum, there is the formation of a reverted phase that overturns a situation that would otherwise result in a decrease of entropy. This is precisely the reason for the spontaneous formation of a precipitate (in proportion to the excess particles). The formation of a precipitate provides a constant multiplicity ratio for every addition of particles beyond the watershed maximum, and the precipitate is therefore in equilibrium with the previously existing liquid phase since they have the same chemical potential. They can then coexist. At equilibrium, the chemical potential of the solution and precipitate are equal, otherwise particles would move from one to the other to maximize the system's entropy. Because of intermolecular forces, the precipitate can also give rise to a hydrated crystal, thereby further increasing the entropy. The further addition of solute particles just increases the proportion of precipitate, but the chemical potential remains equilibrated throughout the system and does not depart from that of the saturated solution; correspondingly, the entropy per parcel of a given volume is constant and uniform. So, at saturation, the further addition of particles results in the formation of a precipitate at constant chemical potential.

IN REAL SYSTEMS, THE CHEMICAL POTENTIAL IS MEASURED IN RELATIVE TERMS

Two questions can arise: (a) How is the chemical potential measured (as it was introduced just because it is easier to be measured than the entropy, at least in relative terms)? and (b) since a transfer of particles to a system can be seen as an accumulation of potentially available energy, how is it that an accumulation of potentially available energy corresponds to an increase of the system entropy?

Regarding the first question, since the chemical potential serves to measure changes in entropy, as it is just defined as a derivative of entropy, it is its relative value that matters, and we can therefore establish its value with some discretionary freedom, according to a reasonable reference scale (but, it needs to be remarked, only to thereby measuring changes in entropy, not its absolute value). In chemistry, the chemical potential is commonly used in reference to the nature of chemical substances, in this case it is employed as if there is a system constituted by a specific amount, typically one mole, of each given chemical species. This way even the changes in the chemical nature of a substance can be measured in terms of chemical potential, as if they were changes of a thermodynamic system, and then even chemical reactions wherein reactants transform into products can be measured in terms of their overall change in the chemical potential, so that their spontaneity can be established. In such instances, the chemical potential is typically dealt with in terms of the Gibbs free energy, which computes the change of the chemical potential consequent to a chemical transformation in the form of equivalent changes of enthalpy (release/uptake of thermal energy plus any work done by volume change) and entropy, based on tabulated standard-state thermodynamic data and equilibrium constants. What can be measured, however, is merely the change in chemical potential relative to the reaction, not the absolute values of chemical potential of the various species. This is because, in chemistry, by not considering the energetic equivalence of matter, and then by ignoring the subatomic state of the system, the chemical potential of any chemical species is necessarily considered only in relative terms. Anyway, the spontaneity of a reaction is directly dependent upon the change that occurs in the process, and not to initial and final values. Relative values of the chemical potentials of the various species can thus be calculated based on the assumption of some reference potential (to provide tabulated standard values that facilitate calculations of changes of the chemical potential in chemical reactions). Indeed, the values of the chemical potentials of the various substances are related to the chemical potentials of the elements they are composed of [42]. As long as we exclude transformations of elements, *i.e.* nuclear reactions, it is therefore useful to refer to the chemical potentials of the elements as reference chemical potentials. Hence, the chemical potential of every element in its common stable state has been agreed to be $\mu = 0$. This is then used as reference potential μ_0. Since in chemical reactions it is not possible to transform one element into another, the values of the various elements themselves are not related to each other [42]. This means that the reference level for any element can be fixed independently (and for the sake of simplicity they have all been fixed to zero), and as in chemical reactions the quantity of each element does not change, this has no effect upon the differences in chemical potential that are being observed [42]. Many compounded substances with negative chemical potential can then be produced from the elements because they are more stable than the elements themselves, at least under some specific conditions [42]. Thus, entropy increases when they are formed and their standard chemical potentials, μ_0, can be calculated from their composition of chemical elements. Any chemical reaction can then be characterized in terms of the overall change in the associated chemical potential. A negative change means the reaction is spontaneous (though if there is a joint transfer of heat and/or of *PV* work then the Gibbs free energy is the proper thermodynamic potential to determine equilibrium conditions, not just the chemical potential). Notably, the net change of chemical potential (or of Gibbs free energy) is independent of the reference potential, μ_0, of each element, and therefore is an objective measure of the spontaneity of the process.

Given the key role that pressure has in the equilibration of open systems (*i.e.*, in the grand canonical ensemble) and that, at a given *T*, the number of particles is proportional to the gas pressure, the chemical potential at every pressure can be calculated in relation to the chemical potential at a reference temperature (that is, the reference chemical potential, μ_0) so that:

$$\mu = \mu_0 + RT \ln(P)$$

The first term, μ_0, is essentially a function of temperature, at least for ideal gases, and it is then the baseline chemical potential at a given temperature, while relative pressure effects are included separately through the second term [9]. However, for real gases the reference chemical potential, μ_0, is different for each gas, and since also the pressure at a given temperature can be different for each gas, the chemical potential at every pressure can be calculated in relation to pressure P_0 (whose reference chemical potential is still μ_0) so that:

$$\mu = \mu_0 + RT \ln(P/P_0)$$

In this case, however, μ_0 is defined at given temperature and also at a reference pressure P_0. In any case, the reference state is assumed to be that of the gas at standard conditions (*i.e.*, 25 °C and 1 atm). It is worthy to note that the concentration of an ideal gas is directly proportional to the system's pressure, thus concentration and pressure can be used interchangeably in calculating the chemical potential of an ideal gas. Pressure is also dependent on the attractive forces that, as we will see, contribute to determine the chemical potential in real gases and is therefore used to measure the chemical potential even for non-ideal gases.

Whereas this notable equation allows us to calculate the reversible changes of entropy in a system at equilibrium, in chemistry the interest is most often focused on the spontaneity of given reactions, that is, on the overall entropy increase associated with such chemical processes. In these conditions, as seen, the Gibbs free energy is typically the thermodynamic potential of choice. What is important is that chemical reactions represent changes, either reversible or not, and therefore what we measure are changes in the Gibbs free energy. We are then not interested in the absolute measures of Gibbs free energy and chemical potential, which are closely linked, as already shown. This means that these changes leave out the baseline chemical potential, μ_0, and the way it is calculated becomes therefore negligible. For chemical reactions, this facilitates (actually, allows) the calculations by setting the values of μ_0 to sensible, but arbitrary, values for the pure elements, as seen. In this way the chemical potential of each substance can be calculated by reference to those values, which are otherwise unknown in absolute terms. This is a very efficacious trick, but great care must be taken in using such values only to measure changes in chemical reactions, which are typically assumed to occur at isobaric pressure (as necessitated by the adoption of the Gibbs free energy).

For real gases, the chemical potential is only measured relative to μ_0, which, differently from ideal gases, is obtained as an arbitrary value, and therefore changes in the chemical potential can only be measured within the conditions used to define μ_0. Clearly, this expression can be used to calculate the chemical potential at any pressure, including pressures higher than the standard conditions, but then a different value of μ_0 must be used, at least for real gases.

The entropy change consequent to a change in the pressure of a system can be much more easily calculated from the volume expansion of the gas corresponding to the exceeding pressure (Fig. **2**). To this aim, on the basis of the ideal gas law $PV = nRT$ rearranged as $P = nRT/V$, the second term of the fundamental equation:

$$dS = PdV/T$$

can be transformed as;

$$dS = nR \cdot dV/V$$

and integrated between two specific values of volume (as entropy is an extensive state function) to yield:

$$\Delta S = nR \cdot \ln(V_2/V_1)$$

The chemical potential of a liquid is given by the same expression used for gases, $\mu = \mu_0 + RT \ln(P/P_0)$, wherein the vapour pressure of the liquid is considered (again, μ_0 is the chemical potential of the pure substance, now a liquid, at standard conditions). In fact, at equilibrium, the chemical potential of a substance present as vapour is equal to its chemical potential in the liquid. Henceforth, as the vapour pressure of every substance in a solution must be at equilibrium with the chemical potential in the solution, as for gases even for liquid systems the vapour pressure can be used to calculate the chemical potentials of all the substances present in the solution [24], at least if they are volatile enough to be measured. In many cases, the solute does not contribute in the vapour pressure of the solution, but it does reduce the effective concentration of the solvent and therefore its vapour pressure. This allows the existence of a close relationship among some characteristics of the system, called colligative properties (dew point depression, which is more directly consequent to the vapour pressure lowering; boiling point elevation, which is a direct consequent to the vapour pressure lowering; freezing point depression; and increased osmotic pressure), that are interlinked because of the effects from the various components on the overall entropy of the solution, which are typically computed in terms of mole fractions [25]. For non-volatile solutes, and in general for every substance present in a mixture, the chemical potential of that substance is:

$$\mu = \mu_0 + RT \ln a$$

where a is the activity of the substance, that is, a measure of the "thermodynamically effective concentration" of that species in the mixture, in the sense that the activity of a real solution effects the species' chemical potential and the colligative properties derived from it, in the same way concentration effects the chemical potential of an ideal solution (that is, a solution where the interactions between the molecules, whether the molecules are the same or different, are all equal). Thus, to measure the concentration or activity of a non-volatile solute, it is possible to measure the vapour pressure of the solvent instead (or another colligative property), at least if there is only one known solute. It is then possible to calculate the activity of the solute from the change of the solvent vapour pressures with respect to the pure solvent. It needs to be highlighted that, for solutions, the standard condition, and therefore an activity of unity, is defined in different ways between the solvent and the solute; for the former, the standard state with unitary activity is that of pure solvent, on the other hand, the accepted convention for a solute is 1 molal, which for an ideal solute, with activity equal to concentration, corresponds to 1 mole per litre of pure solvent. In standard conditions, the chemical potential of a solvent or solute is equal to μ_0. Again, it is reasonable to define μ_0 with different arbitrary conditions for different substances as they do not affect changes in μ.

OTHER FEATURES OF THE CHEMICAL POTENTIAL

It is worthy to note that, in the case of a solution, the chemical potential of the solvent adds up to that of the solute in establishing the system's overall chemical potential [25]. So, since adding some solute particles to an ideal solution means that an equivalent number of solvent particles must be removed to keep the volume constant (or, actually, the mole ratio between them changes in each volume unit), the chemical potential of a component of a mixture cannot change independently of the chemical potentials of the other components [25]. Therefore, if in a binary mixture the less concentrated component is increased, its chemical potential is also increased, but at the same time the more concentrated component is decreased, and its chemical potential is correspondingly decreased by a smaller amount [25]. The reason for the smaller decrease in the chemical potential of the more concentrated component with respect to the less concentrated one is the logarithmic relationship between the chemical potential and concentration: at higher concentrations the change in the chemical potential is smaller than at lower concentrations. In this way, the system's chemical potential and its entropy increase together. In fact, the number of multiset permutations of the spatial partition is augmented and thus the system's positional entropy is increased too, and since, in an ideal solution, there is no change in the momentum and momentum x position entropy components, the system's entropy increases as well. Similar to what happens for a monoatomic gas wherein the chemical potential rises, if particles are added into a given volume, in a non-saturated solution, the addition of solute particles causes an increase of the overall chemical potential (at least, up to saturation), but in this case the increase is lower with respect to a gas because of the replacement of solvent particles needed to keep the volume invariant (or, in general, because of the decrease of solvent concentration). In both instances, however, the system's entropy increases.

This is also true for the relative concentrations of components in binary mixtures for both ideal gases and liquids, wherein the entropy of a system is maximized when the two components have equal molar fractions [25]. In general, equal molar fractions maximize the entropy of ideal mixtures, even if they have more components. As we will see, intermolecular forces can affect this and other important properties of a mixture.

Additional terms can be included in the expression of the chemical potential if the effects of gravity and/or of an electric field (on charged particles) are to be considered, because the latter are constraints that can affect the spatial distribution of particles, and therefore the equilibrium state. For the gravitational contribution we have:

$$-g d(mh)/T$$

that is, gravitational acceleration multiplied by change of either height or particle mass, per temperature. Whereas for the electrical contribution we have:

$$-\mathbb{E}Fdz/T$$

that is, the intensity of the electric field multiplied by Faraday constant multiplied by the change in electric charges, per temperature unit. In these two terms the intensive constant state functions are the gravitational acceleration and the electric field intensity, respectively. In the latter case, the extensive state function is the number of charges (which corresponds to the number of charged particles if they are single-charged); in the presence of an electric field, the higher the number of charges, the more work can be done and therefore entropy is reduced. With regard to the gravitational field, either the mass or the height can represent the extensive state function. When more mass is added to a system, more gravitational potential energy is available, in addition to any effect on the chemical potential and/or the pressure. Whereas, for a given mass, a more elevated height has a similar negative effect on the system entropy[71]. In both cases, more available energy means less entropy.

These two terms are included in the chemical potential when particles are free to move independently, *i.e.* in fluids (wherein these terms participate in the overall equilibration of the chemical potential through the system, like atmospheric gases, which show a gradient of density through different altitudes because the gas particles equilibrate their chemical potentials in consequence to the gravitational field), whereas they are inserted as autonomous terms in the fundamental equation if agglomerated particles, *i.e.* solid bodies, are considered (like suspended particles of silt, which settle out after stirring), because a whole body shows an autonomous equilibration response to gravity and/or to an electromagnetic field, independent of the chemical potential of the system, although not independent of the rest of the system[72]. Rivers and other streams represent a common example of how a water body equilibrates in response to gravity; they preferentially follow the path of least resistance, just like an electric current (which is an example of how the equilibration of electrons to a difference of electric potential acts at a macroscopic level), given a difference in electric potential, distributes itself across different resistive elements in the inverse ratio of the resistance offered by each path (according to the Ohm's law). Analogously, in a "water circuit" water flows in a closed circuit of pipelines, driven by the pressure difference generated by a mechanical pump. In any case, a path of lower resistance will have more matter or energy flowing through it. Therefore, the presence of constraints and restraints will direct the flow, even though the available gradient obviously has an effect as well.[73] In the case of electricity, the difference in electrical potential is relevant in determining the distribution of electric current through available paths, since electrons are negatively charged and they are more attracted to stronger positive net charges. Similarly, pressurized water will flow more intensely toward a lower pressure than a higher one. Anyway, both also flow across a path with lower gradient and higher resistance, if more paths are available; what changes is the intensity of the flow, that is, the time required for equilibration (as flows are changes per time unit), and not the direction of equilibration. In many cases, the kinetic effect can become preponderant in determining the mechanism, *i.e.* the path, of equilibration[74], and this is another way to say that restraints allow for the shepherding, steering, and channelling of spontaneous processes towards a selected direction.

As said, when charged particles distribute singularly in a fluid, the electric term is included in the chemical potential, which is then called the electrochemical potential [7]. This extended form of the chemical potential is very important in the understanding of electrochemical cells (which provide electric current from electrochemical reactions) and electrolytic cells (which employ electric current to perform electrochemical reactions), as well as the energetic reactions involved in respiration and photosynthesis.

SPONTANEOUS CHANGES AND THE CHEMICAL POTENTIAL

Turning now to consider the above-mentioned question regarding the apparently odd fact that an accumulation of potentially available energy corresponds to an increase in the system entropy (when particles are added to a not-too-much concentrated system), it can be envisioned that this is actually a fake question. In fact, an increase of available potential energy in the system is always possible if it is accompanied by a reduction of the overall entropy, and the change of the entropy of the system is not the proper parameter to establish the spontaneity of a transfer (although it is the proper parameter for establishing what happens to the particles once they have entered the system, like an eventual phase separation, as we will see). Indeed, a transfer of particles to a system can occur only if the overall entropy increases, followed by a decrease in the proper thermodynamic potential, aimed at determining equilibrium conditions. As seen, when there is a transfer of particles, additional to energy, the fundamental thermodynamic potential that is minimized at equilibrium is PV/T, that is, pressure is equalized at fixed T and V. So, a transfer of particles is spontaneous if it corresponds to an equalization of pressure (or vapour pressure, for solutions) between the system (or parcel) of origin and the system (or parcel) of destination (which is actually something very intuitive). This maximizes the overall entropy, and, indeed, this latter fact can be directly conceptualized by considering the changes in entropy per transferred particle that occur both at the starting and ending environments; a transfer toward lower particle concentration/pressure means more spatial microstates available per particle and therefore a higher overall entropy [1]. In fact, the overall entropy increases when particles move to a lower chemical potential because the gain of entropy in the lower chemical potential system (or parcel) is higher than the loss of entropy in the higher chemical potential system (or parcel).

Analogously, processes that modify the chemical potential of the system only run "downhill", meaning from higher to lower chemical potentials (at least if no thermal or *PV* effect intervenes to subvert it), because this corresponds to a higher overall entropy (thanks to the greater increase of entropy in the lower chemical potential system), and therefore they are spontaneous only when occurring in such a direction. This is the same thing as transferring heat to a body: its entropy increases notwithstanding a higher temperature is obtained which can be seen as an accumulation of thermal energy, eventually available for operating a heat engine. A direct, discrete heat transfer, however, actually happens only through an irreversible process that increases the overall entropy.

Therefore, as transferring kinetic energy from hot to cold increases the number of thermally accessible microstates in the cold body more than it decreases their number in the hot body, transferring particles from higher to lower pressure/chemical potential increases the number of positional (and position x momentum microstates in the case of gases) accessible microstates in the thinner parcel more than it decreases the number of accessible microstates in the denser one. The presence of parcels with different, unbalanced temperature, pressure, and/or chemical potential would cause a decrease in the overall number of microstates in a system that should hypothetically persist in such an unbalanced macrostate, thus that its microstates' frequencies would be correspondingly altered in the long run. Hence, the Gibbs entropy, by connecting the number of microstates corresponding to a given, or hypothetical, macrostate with the entropy of the studied system, can say whether that system has an equilibrated macrostate or not. In fact, the second law states that in an isolated system, the entropy is maximized at equilibrium and only then is the number of microstates is maximized and *vice versa*, since it is a biunivocal relationship. Thus, the number of microstates is higher at thermal and positional equilibrium than at every state deviating from it; hence, Gibbs entropy is maximal at equilibrium, where it matches the Boltzmann definition of entropy.

INTRODUCING THE RELATION BETWEEN SPONTANEOUS CHANGES AND THE FUNDAMENTAL EQUATION

In the fundamental equation (where E and T, P and V, and μ and N, are the conjugate pairs with respect to changes in the internal energy, since they correspond to diverse kinds of energy transfer), the differential state functions (S, E, V, and N) are all extensive variables, whereas their fixed coefficients are intensive state functions (temperature, pressure, and chemical potential). The intensive state functions may be viewed as potential sources of force, thus any imbalance, or gradient, in these intensive state functions would generate a net force that will induce a discrete change of the extensive state functions to counter the imbalance and thereby restore the equilibrium. For the fundamental equation to be valid (so that changes in the system entropy can be directly calculated from its state functions without having to resort to the complicated calculations of statistical mechanics), any variation of the extensive state function pertinent to each term must occur at a fixed value of the intensive state function(s) included in the same term. That is, to be directly computed, any reversible, or quasi-static, process must occur by varying one (or more) extensive state function at a fixed value of its conjugate intensive state function. Typically, but not necessarily, the intensive state function is fixed by the surroundings, which act as a large buffer for the studied intensive state function during subsequent, gradual equilibrations. In general, to make an entropy change equivalent to a transfer of another state function, we must assume that the process is isothermal, quasi-static, and frictionless, that is, it must be reversible. In order for a process to be reversible (*i.e.*, there is no overall entropy change) then [1]: (a) if heat is transferred, the temperatures must be equal; (b) if volume is transferred, the pressures must be equal; (c) if particles are transferred, the chemical potentials must be equal. These operational conditions assure that the system is at equilibrium. As already remarked, we cannot even numerically define the temperature, pressure, or chemical potential for a system that is not in equilibrium [1]. Nevertheless, in many irreversible isothermal processes, surroundings-based definitions of work and heat [10] can be used to define and measure entropy changes, given that entropy is a state function, and at the end of the process the unconstrained state function of the system equilibrates with that of the surroundings. Clearly, nonequilibrium processes wherein T, P, or μ change in discrete terms, are irreversible, as the overall entropy increases: heat flows towards lower temperature, mobile boundaries move toward lower pressure, and particles move toward lower chemical potential [1].

Although the exact, independent definition of each term of the fundamental equation assumes that the transfer relative to each term occurs at equilibrium conditions, that is, with other parameters being fixed to allow for proper calculation or measurement, discrete changes can be accounted for as well, but, then, their simultaneous effects on the different terms must be considered too. For example, a discrete transfer of particles from a system/parcel toward another system/parcel having a lower chemical potential (that is, a spontaneous diffusion of particles) means that, even supposing the overall temperature is homogeneous, the kinetic energy of these particles is transferred as well. That is, a transfer of heat occurs during the diffusive process [1]. However, unless there is also a change in the molecular interactions among the transferred particles and the new environment, this does not affect the temperature of the systems/parcels and it therefore occurs at thermal equilibrium. In the limit of large systems, wherein the temperature can be considered as representative of the internal energy, an overall increase of entropy can be considered as essentially due to the decrease in the particles' chemical potential as they enter their new environment. On the other hand, if there is a temperature gradient between the systems/parcels, the associated heat transfer must be considered as well. In a spontaneous transfer of either particles, heat, or both, the overall increase of entropy results from the fact that the change of entropy in the origin system/parcel does not match the change of entropy in the destination system/parcel because either the chemical potential, or the temperature, or both, are not equilibrated. In general, an overall increase in entropy corresponds to lost work and the latter is equivalent to the occurrence of waste heat, free expansion, or free diffusion, or a combination of these effects.

CHAPTER 16

Instances of Entropy Change

Abstract: Entropy is maximal at equilibrium. According to the fundamental equation this demands that there is equilibration for every specific interaction term, namely, thermal, mechanical, diffusive, and others. Relevant exemplifications are illustrated for a number of important processes.

Keywords: Chemical potential, Concentration, Diffusive interactions, Earth's life, Electron transport chain, Energy sharing, Equilibration, Hurricane heat engine, Intermolecular potential energy, Katabatic wind, Mechanical interactions, Mixability, Mixtures, Potential wells, Snowflakes, Thermal interactions.

EQUILIBRIUM IS THE REFERENCE FOR SPONTANEOUS CHANGES

The second law requires that "when interacting systems are at equilibrium, their total entropy is maximized" [1]. This demands that there is a complete equilibration between connected systems/parcels regarding their (a) thermal, (b) mechanical, and (c) diffusive interactions [1]. So, approaching equilibrium in the absence of specific constraints, as the changes δQ, dV, and dN can occur independently from each other, then the three terms of the fundamental equation must each individually obey entropy maximization and therefore they must equilibrate independently. Specifically, as two interacting systems/parcels approach equilibrium, three possible processes can occur [1]: "heat flows toward the system with the lower temperature, boundaries move toward the system with the lower pressure, and particles flow toward the system with lower chemical potential". That is, "all changes are in the direction that approaches equilibrium" (because of the net force that results from the overall particles motions), and this guarantees that "interacting systems will converge toward equilibrium rather than diverging away from it" [1]. Therefore, for interacting systems at equilibrium [1]:

Alberto Gianinetti

(a) their temperatures are equal, (b) their pressures are equal, and (c) their chemical potentials are equal. In addition, the macroscopic distribution of matter inside the system is stabilized according to both gravitational and electrical effects. In fact, at equilibrium, there is no net force to induce any further discrete change, and entropy (stability and inertness) is consequently maximized. This, of course, holds even for the different parcels of a system and therefore, for the different phases of an heterogeneous system [32]. In fact, "thermal equilibrium implies that all phases have the same temperature, mechanical equilibrium implies that all phases have the same pressure, and phase equilibrium implies that the chemical potential of every substance has the same value in every phase" [32]. In addition, when either heat, volume, or particles are transferred, with the other two state functions held constant, what typically happens is that [1]: (a) if heat is added, the temperature must rise, (b) if the volume is increased, the pressure must fall, or (c) if particles are added, the chemical potential must rise. Of course, if more than one term could change at the same time, that is, more types of interaction were allowed at once, there could be relevant interferences between these terms; if heat is added to ice, it melts and the temperature does not change; but it also contracts its volume, so it undergoes both thermal and mechanical interactions [1]. Some instances of entropy change occur independently for each term of the fundamental equation, as described below.

INDEPENDENT CHANGES IN THE EXTENSIVE STATE FUNCTION OF EACH TERM OF THE FUNDAMENTAL EQUATION

The first term of the fundamental equation considers the change in energy of the system for purely energetic interactions (*i.e.*, V and N are constant) and, since thermal energy is the component of internal energy transferred in ordinary processes, this term generally correspond to what we define as a heat transfer, δQ (either in terms of conduction, convection, or radiation). In addition to heat, other forms of energy can be transferred. One example that was already mentioned is the instance of light. High-energy radiation from the Sun to the Earth almost equates to the blackbody radiation of the Earth to space, but the two kinds of radiation differ in their entropy content, the former being a more concentrated form of energy, so that the overall entropy increases in consequence to the change. Assuming that the amount of life on the Earth is presently constant (thus that no new overall energy is becoming accumulated in the living organisms), the overall entropy increase occurring in the radiation is partially used to maintain the low level of entropy associated with Earth's life. So, whereas the evolutionary development of photoautotrophic life can be seen as an accumulation of potential energy obtained by exploiting an energy-density difference (partially analogous with the thermal difference between the hot reservoir and the cold reservoir used in a heat engine to produce work), the present dynamic equilibrium can be seen as a process where a local entropy minimum (life) is maintained at the expense of another local entropy minimum (high-energy photons). Hence, no net entropy change needs to occur on the Earth because of existing life, even though an overall entropy increase, through energy transfer from the Sun to the Earth and then to the space, is continuously needed to maintain the local Earth's life entropy minimum. A local decrease in entropy can instead occur when photosynthetic life spreads onto some dead environment, just like a global (*i.e.*, on our Earth globe) decrease of entropy was occurring while photoautotrophic life was expanding on the Earth, but was always coupled to an overall increase of entropy[75].

The second term of the fundamental equation shows that a change of volume (which corresponds to some work transfer, unless the process is completely free and irreversible) always affects the number of the possible arrangements of the particles, and therefore entropy. Any discontinuity in the equilibrium between the pressure of the system and of its surroundings corresponds to a loss of the work that could be obtained reversibly (Fig. **1**). This means that part of the potential energy associated with a pressure gradient is lost during the process. In general, this term considers processes wherein a mechanical equilibration occurs. In turn, this means that a macroscopic mechanical movement should correspond to a discrete *PV* work, and then it is only usually employed when the system is contoured by solid walls (either rigid or elastic) and that at least one of these walls moves. Anyway, the presence of walls is not always necessary; for example in the atmosphere, air circulates in large masses, and since it is a very poor conductor of heat, its temperature strongly depends on the altitude, as its pressure decreases with height and therefore, when moving upwards, it expands because of the reduced pressure and cools almost adiabatically. Thus, even though a large mass of air has no walls, it does *PV* work on its surrounding atmosphere, virtually without absorbing heat (because the low thermal conductivity of air prevents relevant transfers of heat during fast atmospheric circulation), so its internal energy and, hence, its temperature decrease. Conversely, descending air currents undergo adiabatic compression heating. In fact, when an air mass descends, for example, in a katabatic wind flowing downhill over a mountain range, the pressure on the descending air mass increases, its volume decreases, and its temperature increases. Analogously, hurricanes operate as heat engines according to the Carnot's cycle [44]. The "hurricane heat engine" operates between the warm ocean (which acts as a warm reservoir) and the lower temperature at the upper boundary of the troposphere (beyond which the stratosphere acts as a cold reservoir). In the process, part of the heat absorbed at the ocean surface is converted into the mechanical energy of the hurricane wind, while the simultaneous vaporization and condensation of water increases the dissipation of energy to the outer space[76]. As, in addition to the warm air above the oceans, hurricanes can also access and exploit the thermal energy stored in the ocean itself as latent heat of vaporization, they benefit from a larger and more stable thermal reservoir, and therefore are more extended and longer lasting than similar phenomena that occur on the mainland (like tornados).

The last term of the fundamental equation regards diffusive effects, that is, changes of entropy that occur by the transfer of particles, dN, from or to the system. As seen, this effect depends upon the chemical potential, μ, of the system, but since the presence of interactions among the particles most commonly affects it, some additional discussion is required. First, as remarked by Stowe [1], it should noted that "thermal motions cause particles to flow towards regions of lower concentration, simply because there are more particles in the region of higher concentration ready to move out than there are particles out in the region of lower concentration ready to move back in". However, the intermolecular potential energy that is due to the chemical interactions among the molecules in the system must be considered too [1, 4]. In fact, the intermolecular potential energy included in chemical bonds, molecular torsions, electrostatic interactions, and so on [9], represents a constraint on the free and independent movement of the particles[77]. So, "a much stronger attraction between like than between unlike particles favours separation of the components" in multi-components fluids [1].

This effect is thermodynamically relevant when particles entering a new system find a diverse chemical potential, because of their interactions with the new environment. Particles which exhibit a net attraction toward each other will lose potential energy and gain kinetic energy when they "fall" closer. In fact, they have more potential energy when they become farther apart, but more kinetic energy when they are closer together (which is a more stable condition). So, when particles fall into deeper potential wells (with stronger attracting intermolecular forces among particles), the loss in intermolecular potential energy produces a corresponding gain in the thermal energy, and hence a larger accessible number of momentum microstates [1]. In other words, the intermolecular potential energy lost is converted into thermal energy, and then either heat is released to the surroundings or the system's temperature rises. This happens when water freezes, or a fire burns, or concentrated sulphuric acid is mixed with water. Oppositely, a gas with a relevant attraction between its constituent particles decreases its temperature when subjected to free expansion.

CHEMICAL POTENTIAL: CONCENTRATION & INTERACTION EFFECTS

At low temperatures, the molecules move more slowly and spend more time in the preferred orientations with lower potential energy. From gas to liquid to solid state, the molecules come closer together (as a general rule; the higher density of liquid water just above zero with respect to ice is one exception), and then their attracting forces are stronger and the potential energy lower [1]. At higher temperatures, the increased molecular speeds and randomized motions tend to reduce the time molecules spend in such preferred orientations, so the potential wells become shallower [1]. As heat is added to liquids, then not all goes into the motion of the particles, some goes into weakening intermolecular linkages, a sort of work against attracting forces, and thereby raising the potential energy of the particles. Thus, liquids have correspondingly larger heat capacities than gases [1] because more thermal energy has to be spent to counteract attracting forces. Conversely, when particles diffuse into regions of higher intermolecular potential energy (weaker attractive chemical interactions), the thermal energy decreases and the temperature falls [1]. In fact, for some mixtures "the attraction between like particles is significantly stronger than the attraction between unlike particles, so the potential wells are shallower when they are mixed" [1]. Shallower potential wells means that the intermolecular potential energy rises upon mixing, resulting in reduced kinetic energy and therefore the system cools [1]. Anyway, stronger attraction between like than between unlike particles favours the separation of the components, because then the potential wells are deeper and thermal energy is released, affording the particles more momentum microstates [1]. Thus, in some cases, two fluids are immiscible, that is, two phases will form when they are mixed (*e.g.,* water and oil, water and air, methanol and hexane). In this case, at equilibrium, the chemical potential is equal in both phases (each phase is composed almost entirely of one component, but contains a small amount of the other component) and it is lower than the chemical potential theoretically obtained if the two fluids were fully mixed [1]. In this case, the mixing of the two fluids is minimal because the entropy would actually decrease if the fluids were to mix completely. Hence, at equilibrium, the chemical potential of each component, and therefore the overall chemical potential, is the same throughout a system, regardless of the presence of different phases [24].

Consequently, the two considerations that govern the particle distribution are that particles tend to move towards regions of lower concentration and lower intermolecular potential energy. Both these features are measured by μ, which then depends on: (i) particle concentration and (ii) depth of the potential well. Lower concentrations and deeper potential wells (*i.e.,* a stronger intermolecular attraction) mean lower (*i.e.,* more negative) values for the chemical potential, and particles tend to flow from regions of higher chemical potential towards regions of lower chemical potential [1]. Clearly, the intermolecular potential energy can strongly affect the physical state of the system: not only does it determine the mixability of solvents, but also the solubility and the phase-state. Indeed, the physical state of a substance depends on the interactions between its particles: molecules experience an attractive interaction upon close approach, typically due to van der Waals forces, which are present in every molecular system [9]. This is precisely the reason for which the state of a substance can change from gaseous to liquid and then to solid. When the particles enter a more condensed state, they fall into deeper potential wells, and therefore a corresponding latent heat (*i.e.,* enthalpy of phase change) is released. This increases the momentum entropy of the system when external cooling would otherwise tend to reduce it.

In many cases, there is a tradeoff between the two components of the chemical potential; for example, in the phase transition between gas and liquid, the kinetic energy tends to make the particles moving independently, while intermolecular potential energy tends to bring them together. Thus, the temperature has a determining role in establishing the balance between the phases, with high temperatures obviously favouring the less condensed phase. Pressure has an opposite effect. The configuration of lowest chemical potential involves the overall minimization of both intermolecular potential energy and particle concentration; a gain in one area may be offset by a reduction in another [1]. So, as remarked by Stowe [1] "in the microscopic world, we may find some particles in regions of higher (intermolecular) potential energy, albeit in correspondingly lower concentrations" (thus that the chemical potential is equilibrated). An example is the water vapour in our atmosphere. The intermolecular potential energy of a water molecule in the liquid phase is much lower than in the air because of the strong interactions between closely neighbouring molecules. However, there is a net diffusion of molecules from liquid water to the atmosphere if the concentration of water vapour is very low, and the transfer continues till the particle concentration is no longer low enough to offset the increase in intermolecular potential energy [1].

Similar considerations also hold for the freezing of water into snowflakes: lowering the temperature makes water vapour to condensate into ice, driving water molecules against a gradient of low concentration in air to a high concentration in ice, and the asymmetric interactions of water dipoles make them arrange in ordered[78] snowflake crystals rather than maintaining the positional randomness they had in the gaseous phase. In this instance, the concentration component (tendency to a uniform and random distribution of particles through space) of the chemical potential is largely overcome by the effect of the intermolecular potential energy. In fact, the arrangement of water molecules into snowflakes causes them to minimize intermolecular potential energy thus heat (due to fusion enthalpy) is released into the environment. During freezing, the release of heat consequent to the change of phase and to the lower intermolecular potential energy of the hexagonal geometry of the ice crystal lattice (water molecules align themselves to maximize attractive forces, and this causes ice to have a very open crystalline structure, thus its volume is slightly larger than that of cold, liquid water) compensates for the degrees of freedom lost in terms of the concentration gradient (*i.e.*, it makes up for the reduction in the positional randomness) and translates into the six-fold radial symmetry of snowflakes. In Lowe's own terms [45]: "Whenever we observe a spontaneous process that creates positional order, we should be alert for the increase in ways of storing energy that accompanies it; often this concomitant process is the easily overlooked flow of heat from a warmer to a colder object". In fact, snowflakes form in response to a cooling atmosphere and thereupon they release heat, slowing the cooling. The formation of snowflakes is just the most probable equilibration of atmospheric water to a change in the ambient temperature, driven by the probabilistic tendency to maximize the overall entropy and then to minimize the chemical potential (assuming the other terms of the entropy change are not affected).

Analogously, the redox potential of an environment can be exploited to store available energy in molecules having a redox potential different from that of the environment, but that at the same time are relatively stable because of the high activation energy required for the equilibration of their potential with that of the environment. For example, the chemical energy of NADH is used in living cells to provisory store and transfer energy that in an aerobic environment can then be converted into ATP *via* the respiratory chain. The latter is an electron transport chain where electrons are transferred along a series of compounds *via* redox reactions that couple the electron transfer with the transfer of protons across a membrane; this mechanism creates an electrochemical proton gradient that, in turn, drives the synthesis of ATP. The final acceptor of electrons in the electron transport chain is molecular oxygen (which is thereby reduced to water). Thus, the respiratory chain exploits the difference in redox potential between reduced NADH and highly-oxidizing, environmental O_2 to generate the highly phosphorylated ATP molecule, which is used to drive a wide variety of organic reactions thanks to its energy-freeing dephosphorylation mechanism. Another electron transport chain is used for extracting energy *via* redox reactions from sunlight in photosynthesis; again, in aerobic photosynthetic organisms, the final acceptor of electrons is the widely-available atmospheric O_2. All these reactions are enzymatically activated to reduce the activation barriers (energy required to generate high-energy reaction intermediates) that prevent spontaneous, nonfunctional transformations of these molecules. In this way, organic reactions in living cells are teleonomically driven to guarantee the functioning of the cells and eventually of the multicellular organism they belong to. The transfer of electrons from reduced donor molecules to the O_2 obviously occurs in a spontaneous direction, given the large concentration of molecular oxygen in the atmosphere, which therefore represents an oxidative environment. This transfer is clearly driven by electrons falling into deeper wells of oxidation potential and then by an overall increase of entropy, though some work is obtained from this processes by the living organisms to guarantee their own survival. Although there is a transfer of electrons, it is difficult to see the activity of an electron transport chain as a spreading function. Since particles with mass are involved, and their energetic and entropic effects are linked to the (electro)chemical potential of the environment, the term of the fundamental equation that considers diffusive interaction is involved; however, it is not as much a diffusion that makes energy (and entropy) available from donor reduced molecules, but the existence of activation barriers and of enzymes that can override them. Clearly molecules must move to come into contact with enzymes, but no overall change in the spatial distribution of these molecules occurs (as these processes are part of the dynamic equilibrium that maintain a cell at an almost stationary state), thus this spreading does not seem to be the right metaphor to conceptualize these processes. On the other hand, the sharing of energy seems pertinent here, since reduced molecules share their higher potential energy (as they are spots of high redox potential in an oxidative environment) with the surroundings.

CHAPTER 17

The Limits of the Spreading and Sharing Metaphor

Abstract: Some phenomena, though representing instances of entropy change, appear to defy the description of entropy as a function of energy spreading and sharing. Despite its great utility such a description is sometimes difficult to apply because a function levelling down available energy not always exactly acts as a spreading and sharing function. The concept of a physical function levelling down available energy is therefore preferable to understand entropy and the second law of thermodynamics because it has a more general value.

Keywords: Basal energy level, Boltzmann factor, Dissipation of discontinuities, Energy levelling function, Free energy, Macrostate probability, Maximal state probability, Minimal ephemeral deviations, Most probable state, Spreading and sharing, System parcels, Tendency to levelling.

The spreading and sharing function [19, 20] is a useful metaphor that has been conveniently used throughout the previous parts of this discussion. It should nonetheless be clear that some spontaneous processes (*e.g.*, the various phenomena of phase separation, like the formation of a precipitate in saturated solutions, the equilibration of a system to a multiphase state, the genesis of snowflakes, as well as self-assembly processes, including the tendency of surfactants to form micelles, or the hybridization of complementary DNA strands into a double-helix structure) are associated with entropy changes that are not linked to an evident spreading and sharing function (at least in the common sense; spreading and sharing of matter and energy intended with an actual variable meaning according to the condition that for each specific instance turns out to maximize the entropy is always true, but this is a circular definition of entropy).

Alberto Gianinetti

So, the latter appears to be a metaphor that is generally good, but not always valid, although there is an implicit vagueness in meaning that allows it to conveniently be adjusted to different circumstances. Nevertheless, "spreading" and "sharing" are likely to be given primarily spatial interpretations [26]. On the other hand, what every process, included the ones eluding to the idea of a spreading and sharing function, spontaneously does, is to tend to equilibrate to the most probable macroscopic state, which is defined by the combination of probabilistic distributions of matter and energy. Our reliance on these distributions is derived from experience and is, in turn, empirically based on the assumption that the microscopic state of a macroscopic system, or parcel, tends to stabilize uniformly at the basal level of energy identified by the state functions of the system (Fig. **3**). In other words, entropy can also be seen as a function that maximizes the state probability by minimizing the ephemeral, local deviations in the energy of a system from the basal level of internal energy. In the same way, this function levels the energy of different systems when they are combined into a new, single one. In this sense, the Boltzmann factor operates as a function for levelling down the energy of any parcel of the system (Fig. **3**) as well as of any combined system, and therefore it is the fundamental operator that assigns a probabilistic weight to each possible microstate in the canonical partition function[79]. In fact, the second law implies that at equilibrium E is minimized throughout the system (at constant S, V, and N; conditions) with respect to any internal perturbations of the system, or unconstrained internal degrees of freedom [9].

On this view, the maintenance of inhomogeneities in an equilibrated or quasi-static system, like the equilibration to a biphasic state of two chemical species or to a less dispersed state of a single species (from gaseous to liquid to solid, depending on the temperature), should be seen as a consequence to the maximization of the entropy (minimization of the deviations from the basal level of internal energy)[80]. In these cases, the minimization of the otherwise large potential energy for a uniform distribution of particles overpowers the increase in the potential energy associated with the inhomogeneities of the system.

Notably, as seen, minimizing the local, ephemeral deviations of the energy from its basal level of internal energy includes the dissipation of macroscopic discontinuities that can arise when a constraint is removed [34] (Fig. **2**), thus this definition provides a straightforward bridge between statistical and classical mechanics. Since the tendency to the most probable state is a tendency for energy levelling (this is how the Boltzmann factor works; (Fig. **3**)), the second law can be seen as saying that "any isolated system spontaneously tends to equilibrate at an uniform level of energy, saved the effect of constraints". In fact, it is just the levelling of the energy associated with a gradient that represents an increase in entropy and the wasting of potential work (intended as available energy) for any irreversible process (Figs. **1** and **2**). Correspondingly, in isolated systems the stability point is found at the maximum value of entropy or, for other kinds of thermodynamic systems, in configurations of minimum free energy, where the relevant free energy function has to be decided based on the state functions that have been hypothesized to be fixed in the studied system [9, 14]. As a matter of fact, more-probable macrostates are associated with lower free energy, whereas less-probable macrostates are associated with higher free energy [14]. Of course, that's true of the corresponding microstates.

It can be further noted that, once it is constrained within the limits of the system, the net effect of the levelling function (as shown in Fig. (**3**)) is levelling the internal energy of the system to its basal level. This is the effect picked up by the Boltzmann entropy in terms relative to the energy of the considered isolated system, that is, such an expression is focused on the equilibration inside the system. However, in absolute terms defined by the canonical partition function, the energy of the whole system is evidently still subjected to a levelling function (Fig. **3**) that can actually operate only if the system is no longer isolated. This highlights why changing the focus of the expression used to calculate the entropy also changes its calculated value. Actually, the Boltzmann entropy could be better understood by saying that $S \propto \ln \Omega$, which highlights that entropy, at equilibrium, is proportional to the number of microstates, *i.e.* it is an extensive property because it depends on the combinatory distribution of the particles in the available space, and, more generally, it is maximized by extending in phase space the region corresponding to the system's macrostate [46]. Indeed, Boltzmann did not even give thought to the possibility of carrying out an exact measurement of the proportionality constant [47], and therefore of the entropy. He rather meant to show entropy increases monotonically as an isolated gas evolves from nonequilibrium towards an equilibrium state [46].

A function levelling down available energy definitely does not always act as a spreading and sharing function, unless the actual meaning of the latter is deliberately adapted to match with the former in every different context. The former definition is therefore preferable to the latter.

CHAPTER 18

Some Special Instances of Entropy Change

Abstract: A few particular phenomena are quite difficult to frame into the fundamental equation, nonetheless they can be interpreted to the light of the general idea of statistical mechanics that any system and any overall change tend to the most probable state, *i.e.*, a state that is microscopically equilibrated and then macroscopically stable.

Keywords: Correlational behaviour, Deassimilation, Distinguishability of particles, Isomers, Most probable state, Natural patterns, Spontaneous separation, Whirl.

Some reactions that increase entropy are difficult to frame in classical terms, although they are well described at the microscopic level by statistical mechanics. For example, a change in the extent to which molecules are distinguishable can be induced in a system [48] so that it increases the number of accessible positional microstates (*i.e.* the number of distinguishable spatial arrangements of the molecules): if, in a thought-experiment, a catalyst was introduced into a system and caused a spontaneous interconversion of the molecules from pure optical isomers into a racemic mixture [48], the reaction is driven by an increase of entropy that can be seen as an expansion of the phase space in terms of more positional coordinates, since the volume becomes available for two different species of particles in place of the single one previously present. Hence, doubling the chemical species doubles the dimensionality of the phase space, at constant total N. Since the catalyst[81], for which a negligible direct effect on the system entropy should be assumed, is expected to diffuse through the system, then the increase of entropy could be seen as a particular case of the second term of the fundamental expression for entropy changes, wherein an isobaric filling of the available volume occurs by the partial "deassimilation" [48], or appearance, of the new species from the original one, rather than by a more obvious increase in the volume. More exactly, it is the hyper-volume that increases; it depends on the greater combinatory distribution of the particles in the available volume.

Although in chemistry the studied particles are typically atoms and molecules, in high-temperature physics nuclear reactions are considered and the particles are sub-atomic. The highest temperatures occurred at the beginning of the Big Bang, and this made the initial entropy of the newborn universe already quite high. Nevertheless, the entropy of the universe is always increasing since those highest temperatures caused an expansion force that overwhelmed the gravitational force (although it was immense) in a practically adiabatic free expansion (explosion) that is still ongoing and that increases the volume and therefore the entropy of the universe. However, at the same time, the universe cooled off, dampening the entropy increase. The fact that stars and planets formed is just a consequence of the equilibration of matter distribution at the decreased temperatures, which are no longer enough to overcome local gravitational fields. At the same time, with the enormous distances that matter has spread, the overall gravitational effect seems to be too weak to be reverted, just like in a supernova the outer matter is expelled at a speed beyond the escape velocity and gravity cannot pull it back. Notably, the decrease in temperature for the expanding universe has not been caused by work counteracting an external pressure, in fact it is a free expansion, but rather it was consequent to the internal work necessary to overcome the immense gravitational force of its huge mass. In any gravitational field, work has to be done to move a mass outward from the centre of the field (like a weight that has to be lifted up). In this case, the whole mass of the universe was moving outward from its centre owing first to the translational motion of the particles and then to the centrifugal motion of galaxies, into which particles had condensed. So, the particles of the expanding universe gained gravitational potential energy and lost kinetic energy because they used work to drift away from the gravitational centre. In addition, it can be speculated that part of the very initial energy was converted into particle mass to form elementary particles, including quarks and then electrons and protons and, later, atoms and molecules. Since this was a spontaneous phenomenon, given the constraints of sub-atomic forces, it must have represented an increase in entropy as well. In addition, a kind of deassimilation occurred, that is, the appearance of many new species of particles. Even earlier, the second law of thermodynamics possibly selected the number of dimensions in the universe [49].

There are, however, processes that are probabilistically improbable and are therefore associated with negative changes in the overall entropy; for example, a whirl of particles typically does not occurs spontaneously because it would require the synchronous, correlated behaviours for particles to generate a macroscopic effect. After a cup of milk is mixed with a spoon, it spontaneously stops swirling in a relatively short time. Clearly, turbulences tend to dissipate some mechanical energy of the whirl as heat, and the centrifugal force causes a corresponding gradient of milk level in the cup that is levelled down under the force of gravity. For the same reason, in an ideal gas, the particles are expected to randomly deviate from a rotating flux if there is nothing to cause this behaviour. The value of energy to be included in the Boltzmann factor for microstates of this kind of configuration corresponds to the work reversibly obtainable from such a swirling. Something analogous would be the casual, spontaneous separation of hot and cold particles in the cup of milk, to generate a thermal gradient that could be exploited by a heat engine. These types of processes represent highly improbable events corresponding to the transformation of random motion/distribution into correlational, synchronous ones. If they should become macroscopically persistent, this would represent an actual reduction of the number of microstates of the system (maintaining a rotating whirl, or a temperature gradient, would require that some parcels of the system had less possible combinations of either motion direction or kinetic energy with respect to the whole set of combinations actually available if those parcels would have access to all the set of a particle's directionality and kinetic energy present in the system), and therefore to a decrease of overall entropy. As singularities, these phenomena would not be violations of the second law of thermodynamics; this law merely states that they are extremely improbable and ephemeral, so that, in the long run, they must be exceedingly rare and actually unobservable in practice. The assumption of *a priori* equal probability means we feel confident these phenomena are so improbable we can safely keep ignoring them.

Some systems show inhomogeneities that seem improbable, but are maintained by particular interferences, which, if acting systematically, can be considered as a sort of constraint. For example, the presence of sand dunes in deserts or near oceans is sustained by the presence of regular winds. This and other phenomena[82] lead to the spontaneous formation of patterns [50]. They are linked to an increase in entropy, and the systems are usually dynamical and out of equilibrium. The patterns are dissipative, entropy-producing states, driven by an energy flux between some external energy source and heat sink [50], or, more generally, by persisting discontinuities in some state function. In the case of sand dunes, and also sea waves, the energy flux is mediated by wind, and formation of a pattern is caused by the friction between the wind and the wide surface of water or sand; the friction generates air turbulences and these can modify the surface beneath. The surface tends to ripple, and then, interestingly, it becomes more susceptible to interact with the wind. Thus, this appears to be a self-enforcing phenomenon, at least until the wind is lasting and up to the maximum capability of the turbulence to extend the departure from a flat surface (since this latter is the most probable state in rest conditions). In addition, these turbulences turn out to be self-replicating, that is, each one alters the wind-earth interface, thus creating an analogous turbulence, and then an analogous ripple tends to be produced next to it, in the direction the wind blows; this is why these patterns are regular. Indeed, energy flows can cause self-organized patterns to emerge [44]. Both water waves and sand dunes also tend to migrate in the direction the wind blows, consuming wind energy in this macroscopic motion (which is much slower for sand dunes, which also require a recurrent wind to be formed). Indeed, waves and dunes actually are phenomena that dissipate the free energy of the wind. Turbulences and frictions among sand grains produce waste heat. Even the gravitational potential energy that is accumulated as differences in height between the summits and the valleys of waves and dunes is finally dissipated when they reach the end of the sea or of the desert, respectively.

Theoretically, any microstate, or configuration, of a system is considered as a point of the phase space, whose coordinates indicate the positional and energetic values of every particle. The Boltzmann factor weights the probability of each of them. Therefore, this mathematical representation would allow computation of all the effects acting on the probability of a microstate, or configuration, of a system, although at present this would represent an infeasible task. Anyway, it just reflects the general idea that any system, and any overall change, tends to the most probable state, which is a state that is microscopically equilibrated and therefore macroscopically stable. So, the spreading and sharing of energy, as well as other spontaneous processes, such as the conversion of a pure optical isomer into a racemic mixture, and also any improbable event, like a synchronous, correlational behaviour of particles generating a macroscopic whirl of particles, which typically does not occurs spontaneously, are all intuitively understandable as a tendency to, or as deviations from, the most probable state.

CHAPTER 19

Quantization

Abstract: Entropy quantification can be performed under the assumption that both the position of a particle in space and its level of energy can be defined as corresponding to one among many enumerable states, even if their number is hugely high. This means that, if absolute values of entropy have to be computed, neither energy nor space should be continuous variables, even though entropy changes can be calculated in any case. Remarkably, quantum theory just says that's the case, because at a very short scale both energy and space seem to behave like discrete quantities rather than as continuous ones. So, a general string theory, which represents the evolution of quantum theory, appears to be the natural, preferable theoretical framework for the definition of entropy.

Keywords: Absolute entropy values, Continuous distributions, Classical approximation, Enumerability of microstates, Process directionality, Quantum mechanics, Reversible processes, String theory, Wasted work.

A further difficulty in conceptualizing entropy is that the expression for the value of the absolute entropy involves counting microstates through space and energy levels. As seen, the classical approximation to high-energy macroscopic systems that shows continuous distributions of matter and energy provides a valid estimation of the number of microstates supposedly present in the microcanonical ensemble that is theoretically associated with a system. Thus, the value of the classical, continuous canonical partition function is assumed to provide the value of entropy; however, this is strictly true only for the measures of changes in entropy, whereas the absolute value of entropy is only postulated to be that obtained by these calculations [1, 9]. Indeed, in the continuous terms of classical statistical mechanics, the entropy is indeterminate to the extent of an additive constant because of the arbitrariness associated with the calculation of Ω [6].

If we consider that the Gibbs equation, like the Boltzmann equation, requires that the number of microstates is enumerable and finite, whereas the localization of a particle in a continuous space, and hence its velocity, can assume an infinitive series of values (independent of the fact that a measurement of the number of microstates can be in anyway computed by adopting an approximation of choice), it could be deduced that either the actual absolute entropy always has an infinite value (at $T > 0$ K), or that the space, as well as the states of thermal excitation, must be quantized. The latter option is just what is established by quantum physics, and string theory [52] provides an elegant interpretation of space quantization by considering it as a fabric of strings along which quantum energy waves spread (or clump in wavepackets, *i.e.* particles of matter), embedded in an empty nothingness (which is continuous and therefore inaccessible to quantized matter and energy). As remarked by Fermi [6], in a statistical theory based consistently on the quantum theory, all indeterminacy in the definition of Ω and therefore in the definition of entropy, disappears. So, a general string theory is expected to provide an objective quantization of space and then it would be the preferable framework for the definition of entropy. Indeed, the Gibbs entropy has extended to the field of quantum mechanics with the formulation:

$$S = -\mathrm{tr}(\rho \ln \rho)$$

known as the von Neumann entropy [53], which is essentially equivalent to Gibbs entropy, but introduces the density matrix, ρ, characterizing the quantum state of the system in the place of the classical probability, $\wp_m$, and quantifies the degree of mixing throughout the quantum states [53].

As mentioned, even if in classical thermodynamics we cannot have an absolute entropy, thus any measurement of the number of microstates involves an approximation of choice and therefore any calculated value of entropy includes an unknown reference entropy value (which is commonly assumed to be zero by convention, for the sake of simplicity), still we can compare relative numbers and hence obtain valid ratios between the number of microstates and therefore actual entropy differences. For example, if a system with volume V and containing a single particle is expanded to $2V$, the number of microstates doubles such that the entropy change of the system is [9]:

$$\Delta S = k_B \ln(2V/V) = k_B \ln 2$$

So, even if in quantum mechanics we can redefine the entropy to obtain an absolute value for this variable, no physical changes in the equilibrium behaviour of the studied systems would follow, nor would a single physical property or behaviour be affected. Indeed, there is no theoretical constraint preventing a definition for entropy that includes a nonzero value for the unknown reference entropy value [9].

It is worth noting that the quantization of matter, energy, and space also has an effect on the conceptualization of reversibility. Since, in reality, any transfer of matter or energy happens in tiny but discrete amounts (*e.g.*, one molecule at a time; (Fig. **2**)), rather than by mathematical infinitesimals, therefore any real equilibrium takes place in a finite time and even a truly reversible process can occur in a finite time (the infinite number of steps is just a mathematical exigency, since real numbers are not quantized). What really distinguishes reversible processes from irreversible ones is the directionality of the process: indefinite for the former and definite in the latter (and, of course, an overall change of entropy equal or greater than zero, respectively). Therefore, any process that has a direction involves some wasting of work. This is why two additional, equivalent versions of the second law are that "no real process can occur with a 100% efficiency", and that "any real process involves some transformation that reduces the overall available energy, that is, the potential work". Of course, both refer to a real process intended as something that has a given, spontaneous direction; any transformation that occurs at equilibrium conditions does not have a spontaneous direction and is therefore perfectly efficient (though doing nothing that is utilizable).

CHAPTER 20

The Role of Probability in Defining Entropy

Abstract: As a probabilistic law, the second law of thermodynamics needs to be conceptualized in terms of the probabilities of events occurring at the microscopic level. This determines the probability of occurrence for macroscopic phenomena. For the best comprehension of this approach, it is necessary to distinguish between "probabilities", which are subjective predictions of an expected outcome, and "frequencies", which are objective observations of that outcome. This distinction is of help to unravel some ambiguities in the interpretation of the second law of thermodynamics.

Keywords: Boltzmann factor, Equal probability, Equiprobability, Frequencies, Gibbs entropy, Isofrequency, Knowledge subjectivity, Maximum entropy, Observation of facts, Predictions, Probabilities, Quantized space, Subject's uncertainty, Temporal spreading.

Defining the tendency for an overall increase of entropy as the spontaneous tendency to the most probable macrostate implies, specifically, the adoption of the concept of probability in the explanation of a physical property of the universe. Since explaining is commonly intended as an effort to reduce a difficult concept to more primitive, elementary concepts, it is imperative that the idea of "probability" be adequately clarified, since, unfortunately, there is even some debate about its nature. In fact, there are at least two different ways to look at "probability": either as an objective feature of reality, or, rather, as a subjective formulation of the subject's uncertainty about reality [8, 43]. The former view is typically adopted by frequentists, who see probability as an actualized expression of the relative frequency over time of each element in a set of possible outcomes

Alberto Gianinetti

occurring in a given experiment, holding the overall conditions unchanged. It is immediately evident that, if the experiment is what happens in a closed system at equilibrium, probability is the relative frequency over time of each microstate out of the possible ensemble of microstates of that system. As for frequentism, however, a practical hindrance occurs immediately: frequencies can randomly fluctuate around a limit value, thus they are trustworthy only when they attain that limit, that is, only after an infinitely long time of observation. This means that we can be absolutely sure about the actual frequency of an outcome in our experiment/system at the end of the universe, perhaps. It appears to be a quite impracticable approach. Nonetheless, if we have an adequate knowledge of the experiment/system (typically including some observation of frequencies over a discrete time and a reasonable theoretical explanation of what's going on), we can feel able to predict the frequencies of occurrence for the different outcomes (microstates) over time. In other words, we feel we can actualize frequencies over time as present probabilities. This, yet, cannot be obtained without a price. Whereas frequencies would be physical events ascertainable by every competent observer, that is, they are objective facts, probabilities are a prediction of those still unobserved frequencies based on theoretical considerations. They are therefore valid estimations of frequencies if and only if our assumedly adequate knowledge of the experiment/system, and our theoretical considerations based on such knowledge, are really adequate to anticipate them[83]. It is easily recognizable that we cannot be absolutely sure if our knowledge is adequate; in fact, knowledge of experiments/systems widely varies among subjects and some of them can think their own knowledge is adequate while others think the former subjects are totally wrong for trusting in their own knowledge. In other words, knowledge is clearly subjective and certainty is even more so (this is especially evident for the correctness of theoretical considerations). Therefore, frequencies are observable facts about an experiment/system assumed as independent of any observer (apart from considering the Heisenberg's uncertainty principle when they are observed/measured), but probabilities are predictions and are therefore dependent on the experiment/system as much as on the predicting subject. By moving from frequencies to probabilities we gain the possibility of evaluating the experiment/system now, but we give up the full objectivity of the observations. In other words, we assume the risk of being wrong, inexact, or approximate.

However, it is an objective fact that probabilities provide a representation of the reality that is as much objective[84] as the assumptions we used to calculate them do correspond to reality and are exhaustive.

One of the most important assumptions we do in calculating the entropy is assuming the Boltzmann factor truly represents the probability of the corresponding microstate (or configuration) and in any circumstance, that is, either when spatial, thermal, concentration, or intermolecular effects are involved. That all these different aspects can be brought back to the same probabilistic term is a striking feature of our universe. At least, we feel very confident that it is.

As noted above, the subjective formulation sees probability as the subject's uncertainty about reality. Specifically, in statistical thermodynamics, assigning probabilities to the microstates seems not to be really linked with our uncertainty about the real occurrence of a specific microstate at present or at some future, or past, time, but it rather appears more a method to cope with uncertainty in the frequencies of microstates over time. Anyway, uncertainty about states through time (we feel it particularly for future time) is always a feature that characterizes probability. In fact, the uncertainty regarding a sequence of events occurring along time and the uncertainty about an event of the sequence occurring at a given future time are two equivalent instances: the uncertainty about what event occurs at each time is just what renders it impossible to know the sequence of events along time. This is always true, whether the universe is deterministic (that is, the state of the universe along time is fully determined by its initial state) or not. If it is deterministic, then our uncertainty about the exact state of a system at a given time means that we merely do not know, or we are not able to calculate, the exact state of the system and its evolution; if the universe is not deterministic, our uncertainty means that the change in the state of the system through time cannot be calculated at all.

Ultimately, a fully objective definition of the tendency towards an overall increase of entropy, and therefore to an equilibrium state, is "the spontaneous tendency to fluctuate almost only over isofrequent microstates"[85], that is, an isolated system is eventually moved to equilibrium by not-yet balanced forces, and then it essentially fluctuates over microstates that have very similar frequencies of occurrence, and it almost never deviates from these dominating microstates. Even when an isolated system happens to deviate from the dominating configuration of microstates, any deviation tends to be tiny and ephemeral, thus along time the system maintains a distribution of energy and matter that assures its energy is uniformly kept at its minimum level, defined by its macroscopic state functions, and then none of it can be available for doing work, *i.e.* there is no free energy. Conjointly, the number of accessible microstates, that is, the system entropy, is maximized. However, since isofrequency cannot be decided in a finite time, we can conveniently maintain the reference to the "most probable macrostate", assuming that our knowledge, specifically including our assumptions, is adequate in providing a reasonable estimation of the probabilities of its outcomes/microstates, intended as expected, actualized frequencies through time. This can be further checked through numerous finite observations of the distribution of energy and matter in the system, which provide a continuous falsification experiment for the assumption that, along time, no measurable deviations from the theoretically dominating microstates occur. Typically, all these observations fail to contradict our assumption that the energy of an isolated, large system stays at its minimum level, defined by its macroscopic state functions. Therefore, experience corroborates this assumption.

The frequencies of microstates are always relative, in the sense that they sum up to one, and then each microstate has frequency $1/\Omega$ (according to an equal *a priori* probability), as they are relative to what occurs in the specifically studied system. On the other hand, probabilities can be either theoretical estimations of these frequencies or estimations of the overall, absolute probabilities of occurrence, and then they are weighted by the Boltzmann factor. Hence, the Boltzmann and Gibbs entropies consider the probabilities, or frequencies, relative to the studied system, that is, they use values that are normalized to the overall internal energy of the system, so as to consider only the number of microstates. Whereas the canonical partition function, by considering the probabilistic weight (*i.e.* the Boltzmann factor) of the internal energy of the system, actually provides a weighted count of microstates that might be used to calculate an absolute value of entropy, since the partition functions are equations that normalize the probabilities of microstates so that they sum up to one over all microstates (although, presently, any measurement of the number of microstates involves an approximation of choice and therefore any value of entropy is calculated plus an unknown reference entropy value [9]). This is because the relative entropy expressions (by Boltzmann and Gibbs) are devoted to evaluating the equilibrium of the system, while the canonical partition function provides absolute values for energy equilibration that could serve to make comparisons also among different systems. Clearly, this troublesome aspect will be settled when the quantized nature of space and energy is better understood.

Lastly, it can be noted that the Gibbs' formulation specifically shows entropy is maximal when all the microstates have the same frequency, rather than the same probability, that is, their actual occurrence through time is equal. In fact, in a hypothetical system where the frequencies of the isoenergetic microstates stay different forever, the expected, theoretical probabilities of these microstates would still be equal (unless the background theory should change). We nonetheless commonly speak of probabilities in this context just because a system whose isoenergetic microstates spontaneously persist at equilibrium with different frequencies has never been observed. It is only a theoretical construct that serves to address what is deemed not to actually exist. Since it is not existent, and this is a statement consequent to our general postulated assumption that the Boltzmann factor is the determinant of actual frequency, then we can speak of probability even when considering the Gibbs' formulation. That is, we are so confident about the theoretical foundations of the principle of equal *a priori* probability that we consider the expected probabilities to be true estimations of real frequencies, neglecting to distinguish between them. Thus, the Gibbs' expression shows that, when all the accessible microstates of a system have the same frequency, entropy is maximal for that system, given its constraints. This is the very same condition assumed to describe the equilibrium in terms of the Boltzmann entropy (exactly for an isolated system, very closely for an isothermal system). It can now be pointed out that the so-called temporal spreading of the system's energy among microstates over time [20, 36], is just a confirmation, or a re-assessment, in terms of observed frequencies (*i.e.*, according to the Gibbs entropy), of what the principle of equal *a priori* probability already stated in terms of probabilities (and based on which the Boltzmann entropy is assumed to represent the microscopic definition of equilibrium). Hence, the Gibbs entropy is merely a quantitative formulation of the second law of thermodynamics, based on the assumption of equal probabilities for equilibrium microstates.

CHAPTER 21

Outlines of a Verbal Account of the Thermodynamic Entropy for a Pedagogical Approach

Abstract: Starting from the observation of spontaneous phenomena, it can be envisioned that, with time, every isolated system tends to settle into the most equilibrated, stable, and inert condition. In the very long term, this is the most probable state of a system. This can be shown to be a universal law, the second law of thermodynamics, defined as "the tendency to the most probable state". Thereafter, it is intuitive that "a function that measures the equilibration, stability, and inertness of a system" is maximized by the second law. This function is called entropy.

Keywords: Boltzmann factor, Deviation from equilibrium, Dissipation of energy, Energy conservation, Entropy, Free energy, Intensive state functions, Levelling of energy, Macroscopically available work, Meaning of expressions, Microscopic dynamic equilibration, Microscopic forces, Net statistical force, Particles motions, Potentially available work, Probabilistic distributions, Settlement of systems, Second law, Thermodynamic stability, Waste heat, Wasting of work.

Entropy, among other concepts in thermodynamics, is considered to be an abstract concept that is difficult for novices to grasp [26]. Indeed, students (as well as educated laymen) may have difficulties interpreting physical meaning from mathematical expressions [54]. On this regard, Sears [55] claimed that: "There is no concept in the whole field of physics which is more difficult to understand than is the concept of entropy, nor is there one which is more fundamental". Further-

Alberto Gianinetti

more, though thermodynamics is central to our understanding of physics, chemistry, and biology, in most cases these three disciplines treat the topic in distinctly different ways [56]. Therefore, at present, education research has come to focus on understanding how to better teach issues in thermodynamics, drawing across disciplinary boundaries [56]. A number of papers have addressed the teaching of thermodynamics and statistical mechanics, with some trying to spark more interdisciplinary interaction [56]. However, there is little evidence that the research is making progress in this arena [56].

I suggest that, in every context, the concept of entropy can be best introduced by first noting that with time every isolated system settles into the most equilibrated, stable, and inert condition (because of the overall effect of particle motions that tend to counteract any unbalancing of forces). Therefore, this can be shown to be a universal law, known as the second law of thermodynamics and defined as "the tendency to the most probable state", that is, to a macroscopic state whose actual distribution of matter and energy is maximally probable according to the theoretical probabilistic distributions of matter and energy, and also considering the eventual presence of constraints. The probabilistic distributions of matter and energy represent the direction and determine the intensity of the equilibration forces (they are the drive of the observed tendency), and the motions of the particles enable the equilibration forces (they are the effector of the tendency), and therefore they actually push the system into a state of microscopically dynamic equilibrium that corresponds to a macroscopically stable condition.

Since the Boltzmann factor identifies a universal function of energy levelling (which applies to all the intensive state function, namely temperature, pressure, chemical potential, and so on) that rules the state of every system and connects the microscopic and macroscopic levels of the system, then the most probable state is the most energetically equilibrated, stable, and inert state that is actually accessible at the given conditions. Such a macroscopic state implies the minimization, at the microscopic level of dynamic equilibrium, of the frequencies of the microstates wherein parcels (or subsystems) have energy higher than the basal, equilibrium energy. Emphasis should be given, here, to the fact that the overall level of energy is conserved throughout any process (first law of thermodynamics); what changes is rather the distribution of energy and matter (which, indeed, is a constrained form of energy as well).

The levelling function identified by the Boltzmann factor measures any overall unbalancing of microscopic forces, that is, differences in one or more intensive state functions between systems, as well as analogous inhomogeneities within a system. As any unbalance results in a net statistical force that is potentially available to do some kind of work, the levelling function corresponds to available energy (or free energy, that is, energy that is available, or free, for doing an equivalent amount of work). In fact, any unlevelled distribution of energy (like gradients, discontinuities, or inhomogeneities) represents an unconstrained difference of energy, ΔE, that can be subjected to "extraction", that is, the ΔE is potentially available for doing work and it is then called "available energy" (which, in the most common conditions is quantified as either Helmholtz free energy, Gibbs free energy, or PV/T). This available energy is always and only obtained as a deviation from equilibrium conditions, and therefore it represents, by definition, an imbalance in the microscopic forces, caused by the motions of the particles, that results in a net macroscopic force. If not used, or in proportion to the available energy that is not used, this force determines a spontaneous levelling of the internal energy (*i.e.*, the net force pushes the system toward equilibrium, minimizing itself), in any of its various forms (thermal, mechanical, chemical, and so on), and the previously potentially available energy is thereby dissipated. In other words, it can no longer be "extracted" (although the overall internal energy is, of course, conserved). As said, the motions of the system's particles are ultimately responsible for doing the work or dissipating the energy.

By identifying the causal connection between the macroscopic properties of a system and the microscopic particles that constitute it, the Boltzmann factor represents the key mathematical and conceptual function that links microscopic dynamic equilibration (statistical mechanics) with macroscopic equilibrium and available work (classical thermodynamics and mechanics), and also explains why they exist at all.

Quite often the loss, or dissipation, of available energy, which is a wasting of work, causes an increase in the thermal energy of the particles (typically by friction or turbulence), which diffuses to the surroundings as waste heat (which means energy is conserved, but transformed into a less utilizable form; it is less utilizable just because its distribution is more probable). Concomitantly, this spontaneous levelling of the internal energy causes an evident, macroscopic modification of the system that can be computed in terms of system equilibration, stability, and inertness (the last term just means that there is no remaining energy available for doing work because there is no net force at equilibrium; the system is then thermodynamically stable because no net force is left available to modify its macroscopic state). Thereafter, it is intuitive that the second law implies the maximization of "a function that measures the equilibration, stability, and inertness of a system". This function is called entropy. It derives from the probabilistic equilibration of the forces into the system and therefore it is a probabilistic function. As it features a basic aspect of the state of a system, this function is a state variable. Thus, the tendency towards an overall increase in entropy, and therefore the fact that the most probable state of an isolated system is the one with the highest entropy (*i.e.*, it has no free energy left), are evidently set forth by the second law.

Finally, it is known that thermodynamic entropy is isomorphic to information (Shannon) entropy. This will be the argument of a subsequent work; at the moment it can be said that thermodynamic entropy can be fully described without the needing to refer to information entropy.

NOTES

[1]I must point out that the purpose of this book is not to provide a comprehensive discussion of entropy, but rather to illustrate the concept of entropy in its different expressions. Thus, many aspects of thermodynamic physics, like the diverse thermodynamic behaviour of fermions and bosons studied by quantum statistics, are not addressed here. For these topics the reader can refer to a more comprehensive text like [1]. Throughout this book, particles are generally considered from the point of view of chemistry, although some observations on their physical and sub-atomic constitution are occasionally mentioned for illustrative purposes.

[2]In the present work, when a change of entropy is discussed (and eventually designated as dS or ΔS), if not otherwise specified it is meant to refer to the entropy change of the system. When needed, this has been made explicit by

adopting the subscript "sys", like dS_{sys}, whereas the entropy changes occurring in the surroundings and in the entire universe are specified as dS_{sur} and dS_{tot}, respectively. The entropy change of the entire universe is also referred to as either the "overall entropy change" or the "total entropy change" and is meant as the overall, net change of entropy that is consequent to the considered process. Hence, it is calculated as the sum of the entropy changes in the system and its surroundings. This assumes that: (i) only the effects of the studied process are considered (as if the studied process were the only one to occur in the universe) and (ii) the entropy changes in the system and its surroundings are additive and can sum to establish the theoretical "entropy change of the universe" associated with the process. The latter assumption implies that either the system and its surroundings represent an exhaustive partition of the universe or, what is commonly assumed, by "surroundings" we mean all those parts of the universe that can interact with the system and whose entropy can therefore be affected by the observed process. For the purposes of this present introductory work, the two implications are considered as equivalent.

[3]Although the effect of heating, and the associated increase in temperature, on the entropy of a system cannot be measured directly by the heat to temperature ratio pointed out by the Clausius' equation when the process is not isothermal since T in the Clausius' equation would not be constant, the entropy change due to heating can be measured indirectly, for example, by an adiabatic reversible expansion (described below) during which the temperature of the system slowly decreases: the difference of the system's entropy at two temperatures can be measured in terms of volume change (as we will see).

[4]Think of a perfectly elastic ball that falls on a perfectly elastic surface (in an open space) from a given height: the potential energy of the falling ball can be exactly measured in discrete terms and, once the ball comes to rest, such potential energy is completely transformed into waste heat, that is, heat that diffuses to the universe (as there are no insulating barriers). In this case, it is typically assumed that the waste heat produces no temperature increase and therefore it represents a discrete increment of the entropy of the universe exactly equivalent to the released heat.

[5]The universal gas constant is the constant of proportionality that happens to relate the internal energy present in a mole of particles of an ideal gas to the temperature. This relation occurs in the ideal gas law: $PV = nRT$, which can be rearranged as $PV/n = RT$. Since PV has energy units (it is force/area x volume, which can be simplified as force x distance, that is, work) R relates the molar energy of an ideal gas to the temperature.

[6]For real gases, an adiabatic free expansion can also cause a decrease of temperature, since many gaseous molecules show relevant interparticle attractive forces (*i.e.* positive chemical potential energy) that must be overcome to expand the system. This corresponds to internal work that requires energy from the only other source available in normal conditions: the kinetic, or thermal, energy of the gas. This work, and the following temperature drop, however, is no longer equivalent, in terms of entropy change, to the effect of volume change.

[7]The hot reservoir is real in external combustion engines and it is theoretical in internal combustion engines where it actually corresponds to a constant supply of fuel and oxygen mixture.

[8]As remarked by Carnot himself [11]: "The necessary condition of the maximum (work yield) is, then, that in the bodies employed to realize the motive power of heat there should not occur any change of temperature which may not be due to a change of volume. Reciprocally, every time that this condition is fulfilled the maximum will be attained. This principle should never be lost sight of in the construction of a heat engine; it is its fundamental basis. If it cannot be strictly observed, it should at least be departed from as little as possible".

[9]The change of volume that occurs in the cycle affects the efficiency of the engine as well and it must be closely linked to the change in temperature to make the cycle reversible; both the adiabatic compression and adiabatic expansion must be designed so that they leave the system at the same temperature as the hot and cold reservoirs, respectively. Only in this way there is no direct transfer of heat from hot to cold since any change in temperature occurs adiabatically and the two thermal reservoirs assure that, when receiving and doing work, the temperature of the system doesn't change. Since net work is produced because of the difference

between the work done by isothermal expansion at the higher temperature and the work required for the corresponding isothermal compression at the lower temperature, complete conversion of the available thermal energy would be possible only if the gas could undergo an infinite adiabatic expansion, which would maximize the work done (by maximizing the efficacy of the pressure/concentration inside the system at the given T_h) and would thereby bring the temperature to absolute zero and, at the same time, the cold reservoir would exist at a temperature of absolute zero, which would minimize the work consumed by compression (therefore maximizing the conversion of the available thermal energy).

[10]"Extractable" work, *i.e.* available energy, is the maximum fraction of energy that, in a reversible process, can be transformed into work. It is sometimes called "exergy" [16].

[11]Even if the series of equilibrium states appears as macroscopically continuous, it can be thought of either as theoretically infinite, if the integration over infinitesimal steps is used as mathematical approximation to the real process, or finite, if a real quantized transfer of matter or energy is considered for a discrete process.

[12]Indeed, a free expansion of a gaseous system occurs when the latter is allowed to expand into a vacuum. P_{ext} is therefore null throughout the process and no PV work is produced. Consequently, for an ideal gas, the temperature does not change. It can be noted that, during this free process, the pressure of the system drops to zero only if the final volume is infinite; this confirms that using the surroundings-based definitions of work provides a clearer interpretation of the irreversible processes [10].

[13]In this book, "to drive" is intended in the sense of "to steer the movement toward a given direction". In the present context it just means where things tend to spontaneously go.

[14]This reminds us that thermal energy, as well as space, is actually quantized, not infinitesimally continuous.

[15]For a system at $T=0$, all the particles are at a ground state of zero thermal energy, that is, there is only one accessible thermal state. In this case, an increment of temperature equates to both a higher number of accessible states and to a greater spreading among them. According to the Gibbs formula, both contribute to an increase entropy.

[16]Even the application of a magnetic field to paramagnetic material lowers the entropy of the sample and, at a given temperature, the entropy of the sample is lower when the field is on compared to when it is off [24].

[17]By definition, the entropy increases with the logarithm of the overall number of microstates in phase space, which increases proportionally to the temperature, since the variance (spreading) of the Maxwell-Boltzmann distribution of thermal energies is proportional to T. It has to be remarked that, throughout this book, I only refer to macroscopic systems (composed by a huge amount of particles) that are at temperatures far from the absolute zero, so that the quantum effects that affect and complicate the theoretical assessment of the microscopic state of the system at low temperature may be neglected and the high-temperature approximation for classical systems can be adopted [1, 9].

[18]Unless reversibly coupled with an equivalent accumulation of work, which, in other terms, means that if a thermal gradient is levelled to some degree, another gradient, discontinuity, or potential is generated or made steeper to produce a decrement of entropy that exactly balances the entropy increase consequent to the loss of thermal gradient. For this to occur, the heat transfer cannot be direct, but must be mediated by a suitable process aimed to conserve work (as available energy).

[19]It can be noted that the second law of thermodynamics is essentially concerned with changes in entropy rather than with the absolute value of this state function. This is because the second law regards changes, specifically processes, and what characterizes the occurrence of every process is precisely the change of entropy that is associated with it.

[20]If some or all the levels of thermal energy include sublevels of equal energy (these are defined "energy states" and the corresponding energy level is said to be

degenerate) that must be occupied, the number of microstates is reduced accordingly [25] since the different states of an energy level are indistinguishable and they correspond to one and the same microstate.

[21]In the particular case that no energy level is degenerate and the levels are a uniform ladder of the same incremental energy ε, then the molecular partition function reduces to q = $1/(1 - e^{-\beta\epsilon})$ [25].

[22]It can be noted that the momentum partition function already contains all the information expressed by the Maxwell-Boltzmann distribution, but while the former determines how the partition occurs, the latter is expressed in the form of an equation where the independent variable is the energy of each level and the dependent variable is the fractional proportion of particles that are expected to be at that level. As seen, if the levels can include different states that have the same energy, the relative population of each level must take into account the degeneracies of the energy levels [25].

[23]So, in any real heat engine that produces work by cyclical transformations there must always be some waste heat transfer to the environment, in addition to the heat transfer theoretically occurring in a reversible process, to drive the cycle. The desired work can be produced only if the process is irreversible and the overall entropy increases, due to waste heat (or, more generally, to some potentially available reversible work being lost), which serves this purpose. Indeed, one version of the second law of thermodynamics (Kelvin's formulation) is that it is not possible to convert heat completely into work without some other change taking place. The "other change" is typically an increase of entropy, as waste heat or, more generally, as lost work.

[24]To calculate the effects of changing some state functions in a system, other related state functions must be kept constant to allow calculations. Although many combinations are possible, some are most commonly adopted for their wide applications to real systems. Among them, microcanonical conditions are typically considered to occur in isolated systems, wherein the internal energy of the system is conserved and macroscopically unchanged, as stated by the first law of thermodynamics. In fact, E, V and N are assumed to be rigorously constant for

an isolated system, owing to the absence of heat, volume, or mass transfer [9]. Canonical conditions are instead commonly assumed when a closed system thermally interacts with an outside system, or reservoir, or the surrounding environment, at a given temperature. Either isothermal, isobaric, or adiabatic conditions are usually considered when a closed system interacts mechanically with the outside. Finally, grand canonical conditions refer to open systems that can exchange particles and energy, and diffusive phenomena are usually considered at equilibrium chemical potentials.

[25]The symbol "≡" is used to show that two expressions are equivalent and one is the definition of the other.

[26]It may be noted that even assuming a partition unit that includes only one particle (*e.g.,* a cubic space of the size of a particle, or wavepacket), a particle is in no way guaranteed to be centred in the partition unit; it is just assigned to a given unit if the centre of the particle falls into that unit. The exact position of the particle in the unit is not considered and therefore represents a residual, indefinite variable that, however, is deemed not to affect the overall configurational function. In fact, the function assumes randomness in lieu of a systematic arrangement in space and the latter can be observed because of constraints (like a gravitational field) that are typically detectable at multiparticle levels (at the level of groups of more than one particle). Since no systematic effects are known to act at the level of the single particles inside their inherent partition units (*e.g.,* no reasons are known for which the particles should systematically stay centred on the upper left back corner of each cubic unit, and a similar case is expected to be never observed), the arrangement inside the partition unit is assumed to be always random, and therefore the exact position of the particle into the unit is assumed not to influence the entropy of the system. As a consequence, the calculated value of the entropy is always plus an unknown constant that depends on the size of the partition unit adopted for the calculations. There is however, a widely preferred partition, based on the de Broglie wavelength, that will be considered next, and that is derived by assuming a convenient constant that makes the classical canonical partition function agree with the quantum behaviour in the high-temperature limit [9].

[27]Interestingly, the identity of the single particle is not the only criterion of distinguishability. Each of the identical molecules in a crystal lattice, for instance, can still be 'identified' by a set of coordinates, and can therefore be treated as distinguishable because their sites are distinguishable [24]. This is more evident when chemical reactions that are affected by the spatial position of the identical molecules in a crystal lattice are considered: external molecules are much more exposed and prone to react than internal ones, meaning the reactivity of the identical molecules differs according to their position. There is therefore a positional distinguishability of the identical molecules based on their reactivity.

[28]It should be noted that a particle can change its localization throughout space only if it has motion, and it has motion only if it has thermal energy (of which translation motion is a component). Hence, the positional configuration of a free particle depends on the particle, and consequently the system, having some kinetic energy. When initially considering the position in space of particles as the only component of the configurational partition, the latter can be called the positional as well as the translational partition, since the position of the particles changes only because the particles moves. However, the full configurational partition function includes both a positional and a translational, or momentum, component, as we will see.

[29]Roughly speaking, the de Broglie wavelength gives a length scale above which particles will behave classically [9], that is, as spherical and dense masses. For this reason, it is a convenient partition unit and is most largely adopted. Also, it can be obtained analytically from the classical (continuous) approach, which will be considered next.

[30]Hence, the positional partition function increases with the volume of the container, with the mass of the particle (roughly as $m^{3/2}$), and with the temperature (roughly as $T^{3/2}$). It is interesting to note that, according to this equation, the entropy of a system increases as a consequence of an increase in the number of accessible positional states when the volume of the container is increased, as previously stressed, and the molar entropy of a perfect gas of high molar mass is greater than one of low molar mass under the same conditions because the former has more thermally accessible translational states [24].

[31]"Independent" means that the internal energy of the system is the sum of the individual energies of the particles, that is, there are not interactions between the molecules [25].

[32]It can be anticipated that at each instant the ensemble of the particles actualizes as a microstate, identified by a $6N$-dimensional phase space wherein each of the axes corresponds to one of the momentum or position coordinates of each one of the particles, so that the microstates are considered to be points in such a multidimensional phase space.

[33]For a system including N particles, the average number of positional, or translational, states per particle is q/N, and V/N is the volume occupied by a single particle. Therefore, the average separation of the particles is $d = (V/N)^{1/3}$, and it must be much greater than the thermal wavelength for the space partition function to be valid [24].

[34]In this case, the energy of the system (E) is just the sum of the energies of all its particles, thus $E = N \cdot \varepsilon_{mean}$. In addition, under this simplistic assumption, the partition function, defined as the whole set of permutations of the ensemble of N particles, is assumed to be equivalent to the partition function established for the ensemble of all the microstates of the system.

[35]If the particles are indistinguishable, all the states that differ only by a permutation of particles (which amount to $N!$) should be considered as the same state as these states would actually be indistinguishable because they effectively are the same state.

[36]Actually this function, based on the compounding of the many molecular partitions, still has a bias: apart from whether one or another of the indistinguishable particles occupies a given position, there is the fact that only one particle at a time can occupy each position. Thus, the actual number of positions that can be occupied is $V/\Lambda^3\text{-}N$. However, it is commonly assumed that there are far more positions that particles [14], that is, the overall volume of the particles is negligible with respect to the volume of the system, so that $V/\Lambda^3\text{-}N \approx V/\Lambda^3$.

[37]Under these more realistic assumptions, the partition function defined as the

whole set of permutations of the N particles is not equivalent to the partition function established for the ensemble, as some permutations (microstates) are less, or even much less, probable than others.

[38]It can be noted that this is the average over all the accessible microstates of the ensemble for the system. Since the system has just one microstate at each time, the ensemble of accessible microstates is calculated or theorized; however, the microstates are, or can be, real. So, if the term "imaginary" is used to address the states of an ensemble [24], it is only in the sense of "theorized", "predicted", or "actualized" possibilities. Actually, from an epistemological point of view, we could recognize four different ways to see the concept of ensemble (as well as many other concepts): (i) the 'true' ensemble, knowable only through the "eye of God", which would correspond to the set of all the microstates the system would attain through each instant for an infinite time; (ii) a consumptive, experimental ensemble, obtainable (supposedly) by observing the microstates of the system at finite intervals for a long time (say every 0.1 milliseconds for 100 years); (iii) an equivalent way would be observing a very large number of identical systems each at equilibrium at a single instant in time; and (iv) a theoretical ensemble that, based on the present knowledge (assumed to be reliable), allows us to foresee the behaviour of the system and therefore to predict the microstates to which it is capable of acceding. Ensembles are typically considered for systems at equilibrium or in quasi-static conditions, that is, very close to the equilibrium at each instant. A remarkable feature of equilibrium, in fact, is that it is a condition wherein viewing a single system over a long period of time is equivalent to viewing a very large number of identical systems, each at equilibrium, at a single instant in time, that is, time averages equal ensemble averages at equilibrium [9] and both are deemed to be explainable by the same theoretical considerations. In fact, it is only at equilibrium that the ensemble average (calculated from the theoretical ensemble representing the system) of a (not fixed) state function is independent of time and therefore it is equal to the (long) time average in the thermodynamic system of interest [17]. It is thus assumed that both (ii), (iii), and (iv) can be valid estimations of (i), but (iv) is much easier to obtain and therefore it is the most common way to conceptualize an ensemble.

[39]For example, the molecular partition function of a given particle cannot actually

include among the possible localizations of the particles those that are already occupied by other particles; thus that, the accessible volume of the system is lowered by the volume actually occupied by the particles, that is $N \cdot \Lambda^3$ (but, as seen, this is often assumed to be a negligible bias). In addition, microstates that involve some synchrony between the particles, that is, if the latter would happen to collectively move in the same way at the same time, are less probable proportionally to the intensity of such a synchrony, as we will see next.

[40]With respect to the limits of the integrals, it can be noted that each momentum variable has limits that can range from $-\infty$ to $+\infty$ since there are no physical constraints on them. Thus, these integrals are unbounded. In contrast, the allowed values for the position variables depend on the volume of system and are bounded. It is therefore through the limits of the position integrals that the partition function acquires a dependence on V [9].

[41]It should be noted that if the particles are not spherical atoms but molecules, more coordinates must be considered into the phase space, namely the internal (*e.g.,* bond lengths and angles), positional and orientational coordinates of the N molecules. So, if each molecule has K internal coordinates, 3 positional coordinates, and 3 orientational coordinates, the total number of such coordinates per molecule is M = K + 6, over which the integration is performed; correspondingly, the correct power of h, *i.e.* h^{MN}, has to be adopted [27].

[42]The system fluctuates over microstates, some of which may correspond to increasing levels of randomly generated potential energy. In fact, in a gaseous system, the particles could occasionally show a local synchronous translation and fortuitously concentrate in some part of the system, thus generating a pressure gradient, or, again by chance, many particles with higher levels of kinetic energy could gather somewhere, generating a thermal gradient. In an incompressible liquid only the latter event would be possible and in a crystal lattice both are forbidden.

[43]As already seen, the maximization of Ω corresponds, in an isolated system, to the maximization of its entropy. This implies there are not variations of energy in the whole system (by hypothesis, since it is isolated) and, moreover, there are not

variations of energy between parcels of that same system (otherwise Ω is not maximal, at the given E; because any persistent inhomogeneity would reduce the number of microstates that can be actually accessed).

[44]Recall that when speaking of accessible microstates we can mean two things: (a) microstates that are possible because they correspond to the fixed state functions, and (b) microstates that are much more probable because they are fully equilibrated; here accessibility according to both meanings is required.

[45]Indeed, the canonical ensemble is characterized by a given temperature, implicitly assuming that every canonical microstate is rigorously represented by that temperature and no interactions occur between the position and the thermal energy of the particles inside the system, for any microstate, that is, the temperature is assumed to be uniform throughout the system. Hence, the probability distribution of atomic configurations is independent of the distribution of momenta. The former depends only on the potential energy function and the latter only on the kinetic energy function [9]. This is why, in the canonical ensemble, the positional and kinetic contributions to the partition function are considered to be entirely separable. However, in this way, a relevant aspect of the equilibrium of the canonical ensemble is assumed as given, whereas it should be assessed, at least in real cases. This assumption may be reasonable at equilibrium, since it is common to observe either positional or momentum effects, or both, but not an interaction effect of position x momentum. As an example of simultaneous position and momentum effects (occurring during equilibration), without any interaction between them, the mixing of two liquids can be considered thus [1]: when two miscible fluids are mixed, there is an increase of entropy because the molecules of each fluid gain more accessible positional microstates, but at the same time, there can be an effect on the momenta of the particles of the two fluids if the interaction between the molecules of the two species is different from the interaction within the molecules of each species. In fact, for many mixtures the attraction between like particles is significantly stronger than the attraction between unlike particles, so the intermolecular potential energy (*i.e.* the energy derived by the interactions between the molecules) rises upon mixing, resulting, to keep the internal energy constant, in reduced kinetic energy (the system cools off) and therefore in a smaller number of accessible momentum microstates. In

general, upon mixing real fluids, the positional entropy increases whereas the thermal entropy decreases. If the gain in the former is larger than the loss in the latter, then the second law favours mixing [1]. On the other hand, some quasi-static real systems do show interactions between the position and the momentum of the particles; one example is the thermal gradient in the Earth's interior. In fact, the Earth cools very slowly and the geothermal gradient persists even though the Earth is in a quasi-steady-state [28]. The assumption that no interaction occurs between the position and the thermal energy of the particles is therefore not something that can be taken for granted, especially outside ideal systems at equilibrium conditions. It should therefore be made with caution, acknowledging that, in this way, the classical partition function is simplified to the price of taking for granted what is only probable.

[46]This also assumes that no departures from the Maxwell-Boltzmann distribution of energy among particles do occur, which is deemed to be a reasonable assumption for systems at high temperatures, with large numbers of particles and not too thin densities, *i.e.* when exceedingly frequent collisions occur. This assumption holds even for non-equilibrium conditions, but, again, it is not something absolutely granted but for ideal systems.

[47]$\int^V = \int_x \int_y \int_z$ for N particles, which represents the integration over the system volume of the positional (dx_1 dy_1 dz_1) coordinates of each particle, and then represents a set of $3N$ integrals.

[48]Assuming a normal distribution of the fluctuations.

[49]So, for the macroscopic properties of a system, it does not make a difference which ensemble (microcanonical or canonical) we use to define the system's thermodynamic properties. At the microscopic level, however, there are theoretical differences between these ensembles in terms of the microscopic properties and the fluctuations the system is expected to experience [9].

[50]Pressure, like temperature, is an aggregate statistical property of the system ensemble, as at the microscopic level there is not a property that is homogeneous among the particles even though the system is equilibrated at a fixed pressure. Specifically, pressure is related to the average kinetic energy of the moving

particles and is obtained as the aggregate effect of the particles that, at the boundary of each theoretical or physically enclosed parcel, collide with the outside per area unit and time unit. Like for temperature, only the ensemble average kinetic energy of all particles, or the average kinetic energy of a single molecule over long periods of time, provides a proper estimator of the system pressure.

[51]As the maximization of the overall entropy is what actually identifies equilibration, someone could ask why the minimization of free energy parameters is measured instead, and why these parameters are used at all. The answer is that the overall entropy is given by the change of entropy in both the system and its surroundings, but although the system is picked out as a well-defined parcel of the universe, of which we assume to know all the relevant state functions, the surroundings are not so well-defined as for their extension. Some of the state functions of the surroundings, precisely the ones relevant to the studied interaction with the system, are assumed to be known and constant, but the volume and number of particles are assumed to be very large, possibly infinite, and therefore are not definite (and this is an important drawback, since the entropy is an extensive state function). On the other hand, the change of the overall entropy is considered to occur solely in consequence of the studied interaction between the system and the surroundings, so we are greatly facilitated if we can just use a state function of the system in place of the overall entropy, a state function that is determined on the basis of the other known state functions of the system plus the known change in these state functions that is consequent to the studied interaction (this is why the free energy parameter changes depending on the kind of interaction). A parameter that is exactly correlated, even though inversely, with the overall entropy can then be used (by looking at its minimization) to identify the maximization of the overall entropy, even though the latter is not directly quantifiable.

[52]Clearly, if a system is composed of a liquid and a gas phase, there is an interface whose properties are affected by the two phases. The interface is, however, a parcel itself like every other spatial part of the system. If there are no differences in the temperature, pressure, or chemical potential between the interface and its surroundings, as well as between the different phases or even between parcels of

the same phase, then there cannot be differential effects between them (which could cause local macroscopic processes) and the system is at a uniform equilibrium throughout all its macroscopic parts.

[53]Shortly, spreading and mixing up always occur, as a consequence of particles' motions. In this way, eventual inhomogeneities or differences within the system or between adjacent non-isolated systems are levelled up, if no constraints are present.

[54]Of course, this oversimplification is only a mental exercise didactically aimed to help with the understanding of entropy, since, as seen, neither can the interaction effects always be neglected, nor do there exist two different entropies, as this approach could erroneously seem to suggest [36]. Nonetheless, there are actual instances where focusing only on the positional or thermal component of entropy makes sense. First, as seen, the Boltzmann factor makes isothermal (canonical) conditions practically equivalent to isoergonic (microcanonical) conditions, so that deviations from the predominating configuration are rapidly overcome by particle movements and collisions, and deviations from the Maxwell-Boltzmann distribution or any positional x thermal interactions, can be safely neglected. Second, the isothermal conditions assumed in canonical systems allow us to simplify the configurational integral into the volume-defined form that focuses on the positional arrangement of the particles alone. This means that in a large isothermal system the positional effect of the particles is the prevailing aspect of the entropy. Interestingly, our everyday life environments can be roughly approximated to large isothermal systems, although different systems are compartmented at different temperatures, *e.g.*, a room and our body are thermoregulated at different temperatures.

[55]As already noted, a system, even if at equilibrium, can transitorily attain microstates that are less probable (*e.g.*, where some small gradient is casually generated, although it is then instantaneously reverted). By definition, these microstates are not equilibrium microstates, even though they can occur in a system at equilibrium. Therefore, they are proportionally less probable. So, even to calculate the maximum entropy obtainable at the equilibrium in an isolated system, the Gibbs entropy is a theoretically better measurement, due to eventual

departures from the equilibrium toward less probable microstates. Nonetheless, as repeatedly remarked, these fluctuations are so improbable that neglecting them does not have an appreciable impact on the numerical evaluation of the equilibrium entropy.

[56]The Earth can be approximately considered as such a system, since it receives high-energy electromagnetic radiation from the sun and loses a similar amount of low-energy outgoing infrared radiation.

[57]The Boltzmann entropy can be used as a base reference value, even in a non isolated system because, as seen, at equilibrium the fluctuations of a system at a fixed temperature can be approximated to those of a system at a fixed internal energy, as both are negligible in large macroscopic systems.

[58]Precisely, the mean translational energy of a free molecule is $\varepsilon^t_{mean} = 3/2 \cdot k_B T$ and the mean speed of the molecules is $v_{mean} \propto (T/m)^{1/2}$ [25].

[59]In general, the number of degrees of freedom is the number of independent ways by which a dynamic system is free to vary given a set of fixed parameters that characterize that system. In this sense, by "degrees of freedom", we mean the number of microstates that are accessible by (the predominating configuration of) a system, that is, on which the system is normally free to fluctuate. Any constraint, like molecular bonds and physical barriers, limits the number of accessible microstates by excluding the system from some set of otherwise possible microstates (*i.e.*, it limits the extension of the system in the multidimensional phase space).

[60]Although they can have a rotational spin.

[61]That is, there would be macroscopic parcels that are persistently distinguishable for their thermal levels.

[62]It can be useful to remark that the internal energy of a macroscopic isolated system can be dissipated, that is, transformed into a less utilizable form, but it can never be diminished. So, the inherent spontaneous tendency is not towards a lower level of internal energy, but to a lower proportion of utilizable energy, *i.e.* to a more spread and shared distribution of energy and matter that makes the energy

less utilizable and the universe more stable and probabilistically equilibrated. A constant overall level of energy (subject to the first law of thermodynamics) is a concept that is sharply different from the concept that ephemeral energy fluctuations can occur between a system and its surroundings. Hence, the first law assures that the internal energy cannot change by appearing or disappearing from/into nothing, whereas the second law assures its changes occur according to a probabilistic equilibration.

[63] As many spontaneous processes occur along pathways that are not completely irreversible, that is, many natural processes do produce some work (albeit less than theoretically obtainable by a reversible path), it should be noted that free energy (and therefore work) is widely available in nature, but it is not used purposefully. Only sentient beings do that. Nevertheless, evolution and natural selection have been incredibly efficacious in exploiting naturally available free energy (*i.e.*, to obtain useful work); this is why life exists and we can be here to discuss it.

[64] Indeed, if an arsonist deliberately sets fire to a house, it is legally undisputed that, even though fire burns because the entropy increases, the arsonist has played a material role in causing the fire.

[65] No discussion is done here to explain how life evolved. I only offer a reasonable interpretation about how it keeps going.

[66] It is irreversible because part of the heat is not associated with an accumulation of potential energy (*i.e.*, work) and is therefore waste heat or, more generally, lost work, which corresponds to an overall increase of entropy.

[67] Actually, for ideal gases, what each particle species maximizes is just its spatial spreading [34]. However, in processes where some mixing of real fluids can occur, it is the chemical potential that is minimized, rather than just maximizing the spatial spreading of the particles in consequence to differences in concentrations, as we will see. In both cases, this is mathematically expressed as an expansion of the accessible hyper-volume in the multidimensional phase space.

[68] Correspondingly, a further way to state the second law of thermodynamics is

that as two interacting macroscopic systems approach equilibrium, the changes in the variables of the systems will be such that the overall number of states Ω available to the combined system increases, *i.e.* in the approach to equilibrium $\Delta\Omega$ > 0 [1]. This again implies that the spreading of a macrostate over more microstates equals greater stability. Indeed, since every microstate is represented by a point in the phase space, increasing the number of accessible microstates equates to spreading the macrostate of the system over a larger portion of the phase space. Therefore, the delocalization of the system over the phase space is a way of viewing thermodynamic stability. In fact, because any given macrostate will be made up of a different number of microstates, the probability of observing that specific macrostate (rather than another one that, even with the same fixed state functions, fluctuates over microstates that are much less probable because they show a non-equilibrated repartition of matter and/or energy) is proportional to its number of actually accessed microstates [8]. It has been argued that the larger the number of possible microstates for a system, the more dispersed the energy of the system is over time [36]. Leff [20] calls this a temporal spreading of the system's energy among microstates. Actually, this only means that the assumption of equal *a priori* probability holds, and, therefore, that the filling up of the accessible phase space, as well as any increase of the accessible phase space, translates into greater stability. As seen, these two points merely signify that there is equilibrium in the system and that the equilibrated system is as extended as possible.

[69]An equivalent formulation of the fundamental equation is based on the changes of internal energy (at given S, V, N values, that is, for fluctuations that occur without varying the entropy of the system, a condition typically accomplished at the equilibrium), summarized as $dE = TdS-PdV + \Sigma\mu_i dN_i$. This is a formulation of the first law that interrelates changes in E, S, V, and N [1]. It highlights that both *TS*, *PV*, and *μN* are forms of energy. Notwithstanding the energetic nature of these terms, the natural formulation of the fundamental equation is related to the entropy (*i.e.*, the expression shown in the main text), because the transfers of energy, as well as the associated changes in thermodynamic systems, are always a matter of the second law of thermodynamics, and thence of entropy changes. The first law assures the additivity of the three terms, but it is the second law that

describes how they can change.

[70]As we are going to see, the chemical potential of each substance in a normal system is most commonly negative because it has been set to zero for pure elements: since often the latter do not represent a chemically stable condition of matter, common systems that show lower concentration than that of a pure chemical and are composed of more stable substances have values of chemical potential below zero.

[71]As the potential energy of the whole system is not considered in the fundamental equation, it should be noted that the gravitational term applies to a body when the latter is considered as a part of a larger system within which a change of elevation for that body can be envisioned. Gravity then introduces differential effects upon the different parts, or parcels, of the considered system, which must be included in the fundamental equation to fully understand the equilibration inside the system. In general, gravity acts on every parcel of a body and the gravitational force is always proportional to the mass of each parcel. This means that, for a given volume, it is stronger for a denser material. In a heterogeneous system, the gravitational effects of the different phases sum, and since the greater their sum is the greater the resulting decrease in entropy, a system tends to equilibrate so that the physical disposition of the different phases maximizes the system entropy (or, else, it minimizes the decrease in the system entropy when new matter is introduced; remember, anyway, that this is always a consequent of a change obtained through an overall entropy increase). Maximization of the system entropy with respect to its gravitational component always occurs when the denser material is at the lowest possible height. In the case of solid bodies, their weight adds up to a vertical downward force, since they are unable to rearrange their disposition to maximize entropy. When, instead, fluids are considered, the weight of the overlying fluid determines a vertical gradient of pressure that, differently from solids, acts in every direction, according to the Pascal's principle (namely, a change in pressure at any point in an enclosed fluid is transmitted undiminished to all points in the fluid; in fact, in a fluid, particles are free to move in every direction, and therefore they can transfer their average kinetic energy onto every surface unit). This causes specific effects, like those highlighted by the Archimedes' principle, which states that a body immersed in a fluid is subject to

an upward buoyant force equal to the weight of the displaced fluid. As a consequence, the way a fluid equilibrates, in relation to the gravitational field, with insoluble solids or with another, immiscible, fluid shows specific features; whereas solids do not have the capability to arrange themselves to minimize entropy, fluids do have this capability and this generates a number of phenomena that are displayed everyday in our world. The mechanism by which a displaced fluid exerts an upward buoyant force is consequent to the vertical gradient of pressure that develops in the presence of gravity; in a column of fluid, pressure increases with depth as a result of the weight of the overlying fluid. Thus, the pressure at the bottom of an object submerged in a fluid is greater than at the top of the object. This pressure difference results in a net upwards force on the object. It is as if the submerged object, or droplet of immiscible fluid, were squeezed upward by the fluid pressure towards the lower upper pressure. This is the cause of the buoyancy for ships and other vessels, as well as of oil, in water, and also of the rise in the air of a balloon filled with helium. Any floating solid body, or immiscible fluid, is thus subjected to two opposing forces: the downward gravitational force due to its weight (that is, $m \cdot g = \delta_o \cdot V_o \cdot g$; where δ_o is the density of the object and V_o the volume of the object) and the upward buoyant force due to the displaced fluid. If the former is greater than the latter, the object (solid, or a different, immiscible, fluid) will sink, and otherwise it will float. In the case of an immiscible lighter fluid, it will just overlay the denser fluid, since both have the capacity to minimize the entropy of the system and can change their shapes to reach an overall arrangement that minimizes the gravitational entropy of the system. When, instead, a solid object is immersed in a fluid, it cannot spontaneously modify its shape and it will be the fluid alone to change its shape to minimize gravitational entropy, as much as the constraint of the shape of the solid allows. Specifically, in this latter case, we can derive the physical arrangement of the solid body at equilibrium by considering how it is established. The buoyancy force on an object is equal to the weight of the denser fluid displaced by the object, that is, to the density of the fluid multiplied by the submerged volume of the object multiplied by the gravitational acceleration: $\delta_f \cdot V_{df} \cdot g$ (where δ_f is the density of the fluid and V_{df} is the volume of displaced fluid). At equilibrium, the net force on the object must be zero, and thus $F_{net} = 0 = \delta_o \cdot V_o \cdot g - \delta_f \cdot V_{df} \cdot g$, and therefore $\delta_o \cdot V_o \cdot g = \delta_f \cdot V_{df} \cdot g$. Then, $\delta_o \cdot V_o = \delta_f \cdot V_{df}$ and $V_{df} = V_o \cdot \delta_o / \delta_f$, that is, the submerged volume of a buoyant solid object, which cannot modify its shape into a layer overlaying the denser fluid, rests with a portion of its volume submerged, and such a portion is proportional to the ratio between the densities of the body and the fluid. If the body is two thirds as dense as the fluid, then two thirds of its volume will be submerged, thereby displacing a volume of fluid whose weight is equal to the entire weight of the body.

[72]For example, what position a lump of metallic lithium will assume in a system full of nitrogen depends, in the short run, on the presence of gravity and the relative densities of lithium and nitrogen (as lithium floats on liquid nitrogen and sinks in gaseous nitrogen). Then, wherever the lump is positioned inside the system, its chemical potential will slowly equilibrate with that of the nitrogen because chemical and gravitational equilibrations are independent. In the long run, the equilibration of the chemical potential can affect the position of the lump in the presence of a gravitational field, because the reaction product of lithium and nitrogen (namely, lithium nitride, which will form if the temperature is high enough) has a density greater than that of metallic lithium (that is, lithium nitride sinks even in liquid nitrogen). If the lump is small and jagged enough to have a high surface to volume ratio, with time, the formation of an external layer of lithium nitride can cause a significant change of the apparent density of the lump, thus the latter will remain on the bottom even when the temperature is lowered to liquefy the nitrogen. Hence, the two equilibrations (gravitational and chemical) are independent, but they can interact in establishing a final effect.

[73]It should be noted that the available gradient has an effect in directing the flow if it acts as an attractor, that is, there is a continuous spatial effect of the gradient on what is flowing; otherwise, there is no causal nexus by which the intensity of the gradient can affect the path. In other words, a current flows preferentially toward a lower value of an intensive state function (that is, a wider gradient causes a greater flow), if the latter influences the motion of each single flowing particle along the whole flow. Water flows toward a lower pressure because every single water particle has a greater probability to move toward a contiguous region of lower pressure than toward a higher one. Electricity flows toward lower electric potential because every single electron has a greater probability to move toward a contiguous region of lower negative charge (these are the connections between the

macroscopic and microscopic levels of these phenomena). The presence of a continuous causal nexus through space is therefore necessary for the available gradient to affect the direction of the flow (that is, the greater probability of particles motion along the gradient must not undergo local macroscopic failures). The presence of a field, either gravitational, magnetic, or of other nature, that acts through space is another way to guarantee the continuous nexus. For solids, where particles move conjointly as whole bodies, a rope can work as well. On the other hand, if a gradient does not possess causal spatial continuity, it cannot affect the direction of the flow. If water can flow, in a pipeline or stream, toward two paths that both have equal resistance and ending pressure, as both end up (at the same height) with free falls into lower cisterns, they will have the same flow, even if in one case water ends falling into a cistern which is at a much lower height than the other one. Although the final state is more equilibrated (*i.e.* the entropy increase is greater) when water falls to a lower height in a gravitational field, there is no causal continuity between the water in the pipeline, or stream, and that in the end cistern, and therefore no causal effect by which the flow should be preferentially directed through the path leading to the greater entropy increase. In this case, the lower cistern does not act as an attractor, rather it is the combination (according to Bernoulli's equation) of height and atmospheric pressure at the open ends of the two courses that provides equivalent attractors along the two paths. If, instead of a free fall, water reaches the lower cistern through a closed pipe, then the greater gravitational equilibration can exert an effect and divert water to its path because gravity pulls down the flowing water in the pipe and then the resulting pressure drops to negative values (*i.e.*, a tension is developed), thereby assuring the continuity of the causal nexus that increases the flow toward the lower cistern. This is how siphons work, but only up to the atmospheric pressure, above which cavitation occurs (a tension above atmospheric pressure causes quick production of a gaseous phase from water to re-equilibrate its pressure with that of the atmosphere with which water is in contact, this abates the tension and breaks down the continuity of the pulling force). With a similar mechanism, but reverse direction, sap is pulled up in the xylem conduits of trees and other plants by the atmospheric water demand (through transpiration plants facilitate the equilibration of the chemical potential of water between soil and atmosphere, with the former usually largely exceeding the latter). So, in the presence of a difference of some intensive state function, the parcel of lower value can act as an attractor if there is continuity in the transmission of its attractive force along the path to it. Gravity itself poses a great challenge to our understanding of the physical laws governing the universe; since Newton formulated its empirical law, there has been a lot of puzzling about how gravity acts across huge distances, that is, what is the causal nexus that assures the continuity of a pulling force through empty space. A sophisticated and intriguing answer was proposed by Einstein, who raised the idea that the universe has more dimensions than the ones we can perceive, and it is a distortion of the geometry of space caused by gravitational bodies that sink into additional dimension(s) that causes bodies to "slide" toward wells of lower gravitational potential energy. Gravity can then be seen as a geometric property of space and time, that is, the causal nexus between gravitational bodies is represented by a deformation of space itself around them.

[74]Plants, for example, can exploit their specialized transpiration mechanism, which facilitates the equilibration of the chemical potential of water between soil and atmosphere, just because they are more kinetically efficient in causing the equilibration with respect to the direct equilibration obtained by evaporation of soil moisture, although both paths lead to the same entropy increase.

[75]It can be envisioned that in a number of instances wherein the system does not promptly reach the thermodynamic equilibrium, in addition to kinetic restraints (which are always involved, anyway), there is some specific process that maintains the system in a state of macroscopically-dynamic equilibrium, that is, an equilibrium condition that is characterized by levels of some state functions that differ from those expected for the thermodynamic equilibrium. This is the case, for example, of living organisms, the Earth's geothermal gradient, sand dunes, flowing rivers, a working engine, a hurricane, and so on. As already noticed, concomitant to each of these processes, there is some flow of energy that is associated with an overall increase in entropy. All these processes can therefore be seen as either dissipative processes, since they are ultimately involved in the dissipation of available energy, or as "disequilibrium" processes, that is, processes that locally cause, and maintain, a lower level of entropy (lower than that expected for the thermodynamic equilibrium), although they always take place aside of some energy flow that causes an overall increase of entropy. Like

irreversible processes, disequilibrium processes can be described in terms of thermodynamic flow that is driven by thermodynamic forces [44]. Systems sustained by disequilibrium processes can stay for a long time in a state dynamically stable, but far from the thermodynamic equilibrium and are therefore called "nonequilibrium systems" [44]. Differently from what happens at the thermodynamic equilibrium, the dynamic stability of nonequilibrium systems is not associated with inertness. This is due to the fact that they thrive within a flow, which can always be exploited to do work. Humans are nonequilibrium systems and can indeed do a lot of work! Macroscopically-dynamic equilibria are always linked to some kinetic restraint such that the spontaneous flux of energy that is diverted from the main flow of energy to sustain the local perturbation, or nonequilibrium system, is kinetically balanced by the dissipation of the available energy provisionally stored in the local persistent perturbation. This is the essence of every disequilibrium process. If the two fluxes to/from the perturbation were not balanced, either the perturbation would intensify or it would dissolve. If kinetic restraints are favourably arranged, a perturbation could even grow to divert all the original energy flow. The relative velocities of the two fluxes are therefore crucial to the existence and stability of the local persistent perturbation. As these fluxes are associated with changes in the availability of energy (the energy stored in the local persistent perturbation), they also represent opposing changes in the local entropy, whose velocities must match each other to maintain the local disequilibrium at a quasi-stationary state. The physical chemist Ilya Prigogine was the first to stress the importance of the rate of change of entropy in maintaining macroscopically-dynamic equilibria, and he noticed life is just a state based on them. In fact, living organisms are highly specialized structures, or systems, capable of self-organization and self-replication, which are created and maintained by irreversible processes, and were therefore termed dissipative structures by Prigogine [44]. The kinetics of entropy changes are still an ongoing issue, since the complexity of the macroscopically-dynamic equilibria characterizing the living organisms has yet to be fully unravelled. It is, however, clear that restraints and constraints play key roles in these phenomena, essentially depending on their specific arrangement.

[76]As air is a very poor conductor of heat, the direct transfer of heat from the Earth

surface to the outer space is slow by conduction, and, because of the greenhouse effect, radiation is also poorly effective, thus thermal differences can persist for kinetic reasons. On the other hand, convection is much more effective and is therefore involved in most climatic phenomena, like hurricanes. In a hurricane, a large part of the heat absorbed almost isothermically at the surface of the ocean is used as enthalpy, that is, heat of vaporization of water, in a mechanism that dissipates part of the available energy. In fact, some of this enthalpy is irreversibly dispersed into outer space [44].The moist air rises rapidly along the hurricane's eyewall (also because moist air is less dense than dry air) thereby generating a wide vortex of low pressure at the sea level, and since the atmospheric pressure decreases with altitude, the moist air expands adiabatically and cools. As the temperature drops, the water vapour condenses as rain, releasing the enthalpy of vaporization (latent heat), part of which is lost as radiation to outer space in another approximately isothermal transfer [44]. Then, the cooled rain falls and the cooled air at the higher altitude flows out into the weather system, whereby it finally descends again to Earth' surface in an adiabatic compression of dry air. Thus, the vaporization and condensation of water vapour is a mechanism that increases the convective transport of heat from the oceans to higher altitudes (as the air above oceans is less dense than that at the upper troposphere because it is both warmer and moister) where heat is radiated into outer space, thereby increasing the irreversibility of the process, *i.e.*, reducing the obtained work. If this mechanism did not exist and all the heat was present in the form of higher air temperature, the air currents would then be much more intense [44]; that is, the hurricanes would be more efficient in their devastating work. The release of latent heat by the moisture condensing at the high troposphere increases the degree of irreversibility of the process, and it explains why, given the same kinetic restraints, hurricanes occur on oceans instead of hot, dry deserts, notwithstanding the thermal gradient is even stronger there. However, deserts produce large masses of hot air that contribute to generate tropical cyclones and hurricanes that occur elsewhere.

[77]It has been shown [20] that for most classical systems $S_{\text{ideal}}(T, V, N) \geq S_{\text{nonideal}}(T, V, N)$, because position-dependent forces reduce the degree of spatial spreading and the degree of temporal spreading (number of accessible microstates), since

correlations exist between the positions of interacting molecules, and these molecules are not as free to spread their energy as they are in an ideal gas. This relationship also holds for charged particles in a magnetic field and for lattices of discrete and continuous spins.

[78]The ordering here is nothing other than a macroscopic effect of the molecular geometry of water.

[79]Thus, entropy is not a tendency to increase the number of microstates *per se* (which would seem to contradict the fact that the microstates actually visited by a system are only those of the dominating configuration), rather, the number of the equiprobable microstates of the dominating configuration (corresponding to Boltzmann's Ω) weighted for their probabilities is maximized when all local, ephemeral deviations of the energy in a system from its basal level of internal energy are levelled off. For example, an increase of the number of microstates occurs when a constraint that prevents the access of a gas to a given empty space is removed so that space is merged with the previous volume of the system into a new, combined system, whose newly-defined, basal level of internal energy makes any non-equilibrated concentration gradient a high-energy, improbable deviation that is then promptly eliminated by the random movements of the particles. The increased number of accessible microstates is then a numerical accounting of the merging of the systems, not a drive of the particles' random movements. Hence, the temporal spreading of the system's energy among microstates, which corresponds to a less localized and more dispersed energy for the system [36, 20], is an implicit consequence and not the cause of the entropy increase. In other words, the macroscopic equilibrium state of a system is the balancing of all the forces inside the system and implies the minimization, at the microscopic level, of the frequencies of the microstates with energy higher than the basal level (which would correspond to the creation of a net force). This necessarily involves the maximization of the frequencies of the microstates with energy equal, or very close, to that of the basal level. In turn, according to the interpretation statistical mechanics offers for equilibration at the microscopic level, this is described as the filling up of the multidimensional volume in phase space corresponding to the system's dominating configuration, in terms of accessibility. That is, the system's trajectory in phase space in the long run must

visit all the points of the region representing the accessible microstates of the system with equal frequency [17]. Hence, the temporal spreading of the system's energy among microstates over time is just a re-assessment of the principle of equal *a priori* probability. Once equilibrium conditions exist, the temporal spreading of the system's energy among microstates, which is consequently observed in any system, is due to the equilibrium itself, since the latter is temporally antecedent (as frequencies actualize over time for an existing equilibrium). Apart from establishing the right cause-effect relationship, the two things are however biunivocal.

[80]In these instances, entropy is maximized because of the preponderant effects of the positional x kinetic effect (for the Earth's interior) and for other complex interaction effects (for the separation of phases) in the phase space, which overcome the simple effects of sharing energy (thermal effect) and spreading particles (independent positional effects). For the separation of phases, the lower intermolecular potential energy of arrangements with separate species/states increases the probabilities of configurations with either specific internal(species A) x internal(species B) (the species separate) and internal(species A) x internal(species B) x positional (in the presence of a gravitational field, the separated species with the higher density stays on the bottom) coordinates of the molecules (as seen, internal coordinates are given only for molecules and do not exist for ideal monoatomic gases) in the case of biphasic mixtures, or specific internal (at low temperatures the molecules approach each other because the decrease in the intermolecular potential energy overcomes thermal dispersion) and internal x positional (*e.g.*, solid snowflakes fall while water vapour remains in the atmosphere, because of the gravitational field) coordinates of the molecules in the case of a state transition, thereby overweighting, in both instances, the reduced probabilities associated with uneven positional arrangements.

[81]A catalyst increases the rate of a chemical reaction by reducing the energetic barrier (a thermodynamic constraint called "activation energy") that prevents a system from reaching a lower level of potential energy, so that the random fluctuations in the energy of the system (or of the reactants) that were mostly unable to overcome such a barrier in the absence of the catalyst, can now pass it much more easily, thereby allowing the system to move to the lower minimum of

potential energy (*i.e.* to the chemical product) in a proportion that minimizes the chemical potential and maximizes the entropy.

[82]In chemistry, an example of spontaneous formation of patterns is offered by the Liesegang bands or Liesegang rings, a phenomenon seen in many chemical systems undergoing a reaction-diffusion process where a precipitate is formed at opportune concentrations of the reactants and in the absence of convection [51].

[83]The most notable theoretical consideration in this context is the principle of equal *a priori* probabilities, that is, the assumption that, at equilibrium, in an isolated system (constant E, V, and N), all microstates consistent with the macrostate are equally likely. Specifically, the system is equally likely to be in any of its $\Omega(E, V, N)$ microstates; other microstates, at different values of E for example, cannot be visited by the system [9]. This is true by hypothesis under theoretical microcanonical conditions, but not under canonical conditions, which more realistically reflect what happens in the world. Nevertheless, owing to the absolute actual prevalence of the predominating configuration, the principle of equal *a priori* probabilities still offers an approximation of reality that is very good for large systems. It should be noted that the principle of equal *a priori* probabilities for the actually visited microstates is simply a consequence of assuming that the probability of each and every microstate is inversely bound to its level of energy, and only to this function, through the Boltzmann factor. if this truly basic assumption holds, then it is necessarily true that all microstates with the same energy level also have the same probability. Another assumption, linked to the quantification of entropy rather than to its nature, is the assumption of indistinguishability of the particles; permutations of indistinguishable particles are considered to be ineffective with regards to the mixing function, and therefore they reduce to the same microstate. The number of accessible microstates is calculated accordingly. The role of this assumption is evident, for example, in the case of a system with diatomic molecules of a single chemical element. In this case they are assumed to be indistinguishable. However, if we improve our analytical power and detect that there are different isotopes of that chemical element, the mixing of the isotopes should then be contemplated and the number of distinguishable permutations turns out to be higher than previously supposed, that is, we had underestimated the actual entropy.

[84]A representation is a (mental) picture that a subject builds up to describe reality; it is therefore subjective in nature. However, it is a fact that this mental image can have different degrees of correspondence to reality, hence it also depends on reality, that is, it has an objective component. Although a representation always remains subjective in nature (it is not the same thing as the reality it represents, and it depends on the interpretative capability of the subject; like a movie that can never be a usable representation of reality for a subject that is blind and deaf), how it corresponds to reality can go from that of a fairy tale to that of an accurate documentary.

[85]This objectivist formulation also helps to highlight that there is some explanatory circularity when the definition of the second law is presented in terms of entropy. In fact, the formally most correct definition of entropy is in terms of the extension of the accessible hyper-volume in phase space, but this implies the assumption of equal *a priori* probability, which actually is what the second law states. In fact, to state that the tendency to an overall entropy increase is "the spontaneous tendency to fluctuate almost only over isofrequent microstates" is just affirming the assumption of equal *a priori* probability. Thus, the latter can be considered a version of the second law formulated so as to be capable of introducing the concept of statistical entropy. Hence, as already said, a valid theorization must see the definition of entropy in terms of the second law, not *vice-versa*.

References

[1]   K. Stowe, *An Introduction to Thermodynamics and Statistical Mechanics.* 2nd ed Cambridge University Press: Cambridge, 2007.
[http://dx.doi.org/10.1017/CBO9780511801570]

[2]   R. Battino, S.E. Wood, and A.G. Williamson, "On the importance of ideality", *J. Chem. Educ,* vol. 78, pp. 1364-1368, 2001.
[http://dx.doi.org/10.1021/ed078p1364]

[3]   R. Clausius, *The Mechanical Theory of Heat, with its Applications to the Steam-Engine and to the Physical Properties of Bodies.* John van Voorst: London, 1867.

[4]   C.D. Stoner, "Inquiries into the nature of free energy and entropy in respect to biochemical thermodynamics", *Entropy (Basel),* vol. 2, pp. 106-141, 2000.
[http://dx.doi.org/10.3390/e2030106]

[5]   N.C. Craig, and E.A. Gislason, "First law of thermodynamics; irreversible and reversible processes", *J. Chem. Educ,* vol. 79, pp. 193-200, 2002.
[http://dx.doi.org/10.1021/ed079p193]

[6]   E. Fermi, *Thermodynamics.* Prentice-Hall: New York, 1937.

[7]   E. Keszei, *Chemical Thermodynamics: An Introduction.* Springer Science & Business Media: Heidelberg, 2012.
[http://dx.doi.org/10.1007/978-3-642-19864-9]

[8]   K.A. Dill, and S. Bromberg, *Molecular Driving Forces: Statistical Thermodynamics in Biology, Chemistry, Physics, and Nanoscience.* 2nd ed Garland Science: New York, 2011.

[9]   M.S. Shell, *Thermodynamics and Statistical Mechanics: An Integrated Approach.* Cambridge University Press: Cambridge, 2015.
[http://dx.doi.org/10.1017/CBO9781139028875]

[10]   E.A. Gislason, and N.C. Craig, "Cementing the foundations of thermodynamics: Comparison of system-based and surroundings-based definitions of work and heat", *J. Chem. Thermodyn,* vol. 37, pp. 954-966, 2005.
[http://dx.doi.org/10.1016/j.jct.2004.12.012]

[11]   S. Carnot, *Réflexions sur la puissance motrice du feu et sur les machines propres à développer cette puissance. Bachelier: Paris, 1824. English translation: Reflections on the motive power of heat, and on machines fitted to develop that power.* J. Wiley & Sons: New York, 1897.

[12]   Z.S. Spakovszky, Irreversibility, Entropy Changes, and "Lost Work". A topic (Subsection 6.5) from the course titled "16. Unified: Thermodynamics and Propulsion" taught by Prof. Z.S. Spakovszky at MIT, 2008. [Online]. The online HTML version of full Lecture Notes by E.M. Greitzer, D. Quattrochi, Z.S. Spakovszky, and I.A. Waitz is Available from: http://web.mit.edu/16.unified/www/FALL/ thermodynamics 2008 [Accessed: 15th Jan. 2015].

[13]   E.A. Gislason, and N.C. Craig, "General definitions of work and heat in thermodynamic processes", *J. Chem. Educ,* vol. 64, pp. 660-668, 1987.

[http://dx.doi.org/10.1021/ed064p660]

[14] J.C. Lee, *Thermal Physics: Entropy and Free Energies.* 2nd ed World Scientific Publishing Co.: Singapore, 2011.
[http://dx.doi.org/10.1142/8092]

[15] L.E. Strong, and H.F. Halliwell, "An alternative to free energy for undergraduate instruction", *J. Chem. Educ,* vol. 47, pp. 347-352, 1970.
[http://dx.doi.org/10.1021/ed047p347]

[16] J. Honerkamp, *Statistical Physics: An Advanced Approach with Applications.* 2nd ed Springer: Heidelberg, 2002.
[http://dx.doi.org/10.1007/978-3-662-04763-7]

[17] T.L. Hill, *An Introduction to Statistical Thermodynamics.* Dover Publications: New York, 1986.

[18] J.W. Gibbs, *Elementary Principles in Statistical Mechanics, developed with especial reference to the rational foundation of thermodynamics.* Dover Publications: New York, 1902.
[http://dx.doi.org/10.5962/bhl.title.32624]

[19] H.S. Leff, "Thermodynamic entropy: The spreading and sharing of energy", *Am. J. Phys,* vol. 64, pp. 1261-1271, 1996.
[http://dx.doi.org/10.1119/1.18389]

[20] H.S. Leff, "Entropy, its language, and interpretation", *Found. Phys,* vol. 37, pp. 1744-1766, 2007.
[http://dx.doi.org/10.1007/s10701-007-9163-3]

[21] P. Atkins, *The Laws of Thermodynamics: A Very Short Introduction.* Oxford University Press: Oxford, 2010.
[http://dx.doi.org/10.1093/actrade/9780199572199.001.0001]

[22] W.B. Jensen, "Entropy and Constraint of Motion", *J. Chem. Educ,* vol. 81, p. 639, 2004.
[http://dx.doi.org/10.1021/ed081p639.2]

[23] D.F. Styer, "Insight into entropy", *Am. J. Phys,* vol. 68, pp. 1090-1096, 2000.
[http://dx.doi.org/10.1119/1.1287353]

[24] P. Atkins, and J. dePaula, *Atkins' Physical Chemistry.* 9th ed Oxford University Press: Oxford, 2009.

[25] P. Atkins, J. dePaula, and R. Friedman, *Physical Chemistry: Quanta, Matter, and Change.* 2nd ed Oxford University Press: Oxford, 2014.

[26] F. Jeppsson, J. Haglund, and H. Strömdahl, "Exploiting language in teaching of entropy", *J. Baltic Sci. Educ,* vol. 10, pp. 27-35, 2011.

[27] J. Simons, *An Introduction to Theoretical Chemistry.* Cambridge University Press: Cambridge, 2003.

[28] A.M. Hofmeister, and R.E. Criss, "How irreversible heat transport processes drive Earth's interdependent thermal, structural, and chemical evolution", *Gondwana Res,* vol. 24, pp. 490-500, 2013.
[http://dx.doi.org/10.1016/j.gr.2013.02.009]

[29] P. Atkins, J. dePaula, and R. Friedman, *Quanta, Matter, and Change: A molecular approach to physical chemistry.* W.H. Freeman and Company: New York, 2009.

[30]   F.L. Lambert, "Entropy is simple, qualitatively", *J. Chem. Educ,* vol. 79, pp. 1241-1246, 2002.
[http://dx.doi.org/10.1021/ed079p1241]

[31]   N.C. Craig, "Entropy analyses of four familiar processes", *J. Chem. Educ,* vol. 65, pp. 760-764, 1988.
[http://dx.doi.org/10.1021/ed065p760]

[32]   R.G. Mortimer, *Physical Chemistry.* 3rd ed Elsevier Academic Press: Burlington, 2008.

[33]   H.S. Leff, "Removing the mystery of entropy and thermodynamics – Part II", *Phys. Teach,* vol. 50, pp. 87-90, 2012.
[http://dx.doi.org/10.1119/1.3677281]

[34]   A. Ben-Naim, "An informational-theoretical formulation of the Second Law of Thermodynamics", *J. Chem. Educ,* vol. 86, pp. 99-105, 2009.
[http://dx.doi.org/10.1021/ed086p99]

[35]   J.W. Gibbs, "Graphical methods in the thermodynamics of fluids", *Trans. Conn. Acad,* vol. 2, pp. 309-342, 1873.

[36]   F.L. Lambert, "Configurational entropy revisited", *J. Chem. Educ,* vol. 84, pp. 1548-1550, 2007.
[http://dx.doi.org/10.1021/ed084p1548]

[37]   H.S. Leff, and F.L. Lambert, "Melding two approaches to entropy", *J. Chem. Educ,* vol. 87, p. 143, 2010.
[http://dx.doi.org/10.1021/ed800067a]

[38]   E.N. Economou, *The Physics of Solids: Essentials and Beyond.* Springer: Heidelberg, 2010.
[http://dx.doi.org/10.1007/978-3-642-02069-8]

[39]   E.D. Schneider, and D. Sagan, *Into the Cool: Energy Flow, Thermodynamics, and Life.* University Of Chicago Press: Chicago, 2005.

[40]   R. Baierlein, *Thermal Physics.* Cambridge University Press: Cambridge, 1999.
[http://dx.doi.org/10.1017/CBO9780511840227]

[41]   J.P. Lowe, "Heat-fall and entropy", *J. Chem. Educ,* vol. 59, p. 353, 1982.
[http://dx.doi.org/10.1021/ed059p353]

[42]   G. Job, and R. Rüffler, *Physical Chemistry from a Different Angle - Introducing Chemical Equilibrium, Kinetics and Electrochemistry by Numerous Experiments.* Springer: Heidelberg, 2016.

[43]   E.T. Jaynes, *Probability Theory: The Logic of Science.* Cambridge University Press: Cambridge, 2003.
[http://dx.doi.org/10.1017/CBO9780511790423]

[44]   D. Kondepudi, *Introduction to Modern Thermodynamics.* J. Wiley & Sons: Chichester, 2008.

[45]   J.P. Lowe, "Entropy: Conceptual disorder", *J. Chem. Educ,* vol. 65, pp. 403-406, 1988.
[http://dx.doi.org/10.1021/ed065p403]

[46]   L. Boltzmann, *Wissenschaftliche Abhandlungen.* Barth: Leipzig, 1909. [reissued by Chelsea: New York, 1969]

[47]   M. Planck, "The genesis and present state of development of the quantum theory", In: *Nobel Prize Lecture, June 2, 1920. Nobel Lectures, Physics 1901-1921.* Elsevier Publishing Company: Amsterdam, 1967. [Accessed: 12th Apr. 2016]

[48]   J.N. Spencer, and J.P. Lowe, "Entropy: The effects of distinguishability", *J. Chem. Educ,* vol. 80, pp. 1417-1424, 2003.
[http://dx.doi.org/10.1021/ed080p1417]

[49]   J. Gonzalez-Ayala, R. Cordero and F. Angulo-Brown, "Is the $(3 + 1)-d$ nature of the universe a thermodynamic necessity?", *EPL (Europhysics Letters),* vol. 113, p. 40006, 2016.
[http://dx.doi.org/10.1209/0295-5075/113/40006]

[50]   L. Goehring, "Pattern formation in the geosciences", *Philos. Trans. R. Soc. A,* vol. 371, p. 20120352, 2013.
[http://dx.doi.org/10.1098/rsta.2012.0352]

[51]   H. Nabika, "Liesegang phenomena: spontaneous pattern formation engineered by chemical reactions", *Curr. Phys. Chem,* vol. 5, pp. 5-20, 2015.
[http://dx.doi.org/10.2174/1877946805011509081100839]

[52]   B. Greene, *The Elegant Universe: Superstrings, Hidden Dimensions, and the Quest for the Ultimate Theory.* 2nd ed W.W. Norton & Company: New York, 2010.

[53]   J. Von Neumann, *Mathematical Foundations of Quantum Mechanics.* Princeton University Press: Princeton, 1955.

[54]   K. Bain, A. Moon, M.R. Mack, and M.H. Towns, "A review of research on the teaching and learning of thermodynamics at the university level", *Chem. Educ. Res. Pract,* vol. 15, pp. 320-335, 2014.
[http://dx.doi.org/10.1039/C4RP00011K]

[55]   F.W. Sears, *Principles of Physics I. Mechanics, Heat and Sound.* Addison-Wesley Press: Cambridge, MA, 1944.

[56]   B. Dreyfus, B. Geller, D.E. Meltzer, and V. Sawtelle, "Resource Letter TTSM-1: Teaching thermodynamics and statistical mechanics in introductory physics, chemistry, and biology", *Am. J. Phys,* vol. 83, pp. 5-21, 2015.
[http://dx.doi.org/10.1119/1.4891673]

SUBJECT INDEX

www.ingramcontent.com/pod-product-compliance
Lightning Source LLC
Chambersburg PA
CBHW050845220326
41598CB00006B/433